LUXURY
DESIGN FOR LIVING

Schiffer Publishing Ltd
4880 Lower Valley Road • Atglen, PA 19310

A CELEBRATION OF THE WORLD'S MOST EXQUISITE GOODS

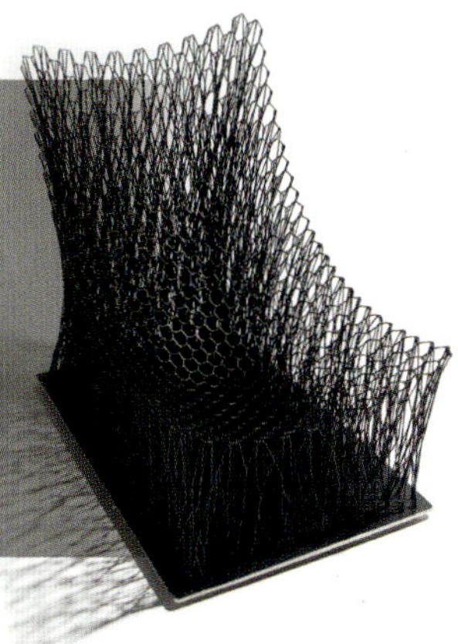

LUXURY

DESIGN FOR LIVING

STEVE HUYTON

Other Schiffer Books by Steve Huyton:

Limited Edition Watches, ISBN 978-0-7643-5164-8

Modern Masters, ISBN 978-0-7643-5384-0

Library of Congress Control Number: 2017935603

Designed by Justin Watkinson
Cover design by Danielle Farmer
Type set in BauerBodni BT/BentonSans

ISBN: 978-0-7643-5421-2
Printed in China

Published by Schiffer Publishing, Ltd.
4880 Lower Valley Road
Atglen, PA 19310
Phone: (610) 593-1777; Fax: (610) 593-2002
E-mail: Info@schifferbooks.com
Web: www.schifferbooks.com

CONTENTS

CHAPTER 1. Architecture

CHAPTER 2. Horology

CHAPTER 3. Transportation

CHAPTER 4. Lifestyle

INTRODUCTION

Like most people, I have always been fascinated by the world of luxury. It is the originality and exclusive nature of ultra-high-end products that intrigue me. As there are usually no budgetary restrictions, designers can employ the world's finest craftsmen and use the most extravagant materials to realize their creative vision. This book is a visual exploration of a broad range of aspirational products.

The book is divided into four sections: Architecture, Automobiles, Horology, and Lifestyle. Each of these categories features twelve companies and three of their most exclusive products. Some of the brands showcased will be instantly familiar, while others are not as widely known. All of the products are examples of the finest cutting-edge design.

1. ARCHITECTURE

AR Design Studio
Hampshire, England

Andy Ramus founded AR Design Studio more than fifteen years ago after working on large-scale projects in London for several years. Each commission, ranging from new builds to innovative additions, demonstrates that great richness can be achieved by creating an environmentally friendly building that focuses on the needs of the client and site.

Abbots Way
HAMPSHIRE, ENGLAND

Abbots Way, completed in 2011, is bordered by mature trees, accessed by a private lane, and overlooks an idyllic small lake. A white box containing the five bedrooms appears to float on the axial stone walls, inspired by a local twelfth-century abbey. The living spaces nestle below it and feature natural stone floors, a state-of-the-art kitchen, and floor-to-ceiling glazing.

4Views

HAMPSTEAD, ENGLAND

4Views, designed in 2012, has spectacular views of Winchester city. Designed around the three Ns—nature, natural light, and natural air—the cedar-clad house consists of four interlocking boxes, and the void at the center contains an open garden with a single olive tree. The upside-down floor plan (bedrooms are on the bottom) features sliding glass panels, a courtyard, large balcony, and huge covered patio that bring the outdoors in.

The Boathouse

COTSWOLDS, ENGLAND

This accessory dwelling serves as a secluded and intimate retreat apart from the owner's main house, a converted eighteenth-century barn. The four-legged steel frame structure sits on concrete piles sunk into the lake bed. Traditional materials such as oak and slate acknowledge the rural setting and combine with concrete and glass to create the perfect haven.

Bates Masi + Architects

Sagaponack, New York

Bates Masi + Architects, owned by Harry Bates and Paul Masi, is an architecture practice with deep roots in New York City and the East End of Long Island dating back more than fifty years. In addition to designing a range of exquisite residential properties, the firm's commissions include schools, offices, hotels, restaurants, retail, and furniture in the United States, Central America, and the Caribbean.

Elizabeth II

AMAGANSETT, NEW YORK

Acoustics drove the design of this house, on a bustling street in a resort town. The house is made from a series of parallel, nearly twenty-inch-thick insulated concrete walls that ascend in height, buffering the street sounds and providing thermal mass and structural support. The basement-to-second-level stair treads also create an acoustical experience. They taper in thickness, changing the sound of footfalls as the residents go up and down the stairs.

Far Pond

SOUTHAMPTON, NEW YORK

Far Pond, built around the framework of a 1970s kit house, overlooks an estuary, bay, and the ocean. The expanded 3,100-square-foot structure uses prefabricated solid steel wall panels that multi-task throughout the design: The steel is perforated over windows and doors, baffling sunlight; fins protruding from the wall panels accept shelving, seating, and countertops, and the steel staircase framework morphs into a desk. This minimizes the need for additional components and their wasteful packaging and shipping.

Mothersill

WATERMILL, NEW YORK

This home was incorporated into a site that contains two historic structures by renowned architect Andrew Geller. Inspired by beachfront construction, the architects used a boardwalk to tie the structures and gardens together. The boardwalk extends through the main house and out onto a cantilevered deck overlooking a sloping wetland and creek. The surface of the path folds up and over to enclose the main house, becoming a wall and roof, thus interweaving the house with the existing landscape.

Charles Wright Architects

North Queensland, Australia

Charles Wright Architects (CWA) is an innovative design practice in idyllic Port Douglas. The business, founded in 2004, has a second office in Melbourne. Wright's training as a sculptor and painter is evident in the sculptural and compositional qualities of his buildings, which are often built on challenging sites and must withstand adverse weather conditions.

Cassowary Estate

NORTH QUEENSLAND, AUSTRALIA

Cassowary Estate is in Daintree on a sixteen-hectare (forty-acre) plot surrounded by a world heritage tropical rainforest. Designed for low environmental impact, a high profile, and sustainable luxury, it consists of discrete indoor-outdoor living pavilions under an unusual "ribbon" roof canopy. This dwelling is designed for cross-ventilation and to adapt to the annual fluctuations in weather. The surrounding wetlands and water features aid in evaporative cooling.

Glass House

QUEENSLAND, AUSTRALIA

The Glass House in Cairns is a reinterpretation of renowned architect Philip Johnson's Glass House in New Canaan, Connecticut, US. Its main living areas are literally open to the outdoors, while bedrooms, bathrooms, a lounge/music room, and office open via sliding glass panels and are individually air-conditioned on an as-needed basis. The central interior courtyard holding the swimming pool acts as a show-stopper "rain curtain" in tropical downpours.

Oak Beach House (ELANORA)

QUEENSLAND, AUSTRALIA

Oak Beach House sits on a rare hilltop site overlooking the Coral Sea and surrounding world heritage rainforest. The entire home opens up to terraces, an outdoor kitchen, and an infinity pool, allowing the inside and outside to blend seamlessly. Interior features include marble-tiled bathrooms, polished concrete flooring, a state-of-the-art kitchen, an integrated sound system, and a commercial-size gym.

Durbach Block Jaggers Architects

Sydney, Australia

The award-winning Durbach Block Jaggers Architects was established by forward-thinking architects Neil Durbach and Camilla Block. In addition to many high-profile commercial projects, they also create beautiful homes.

Holman House

SYDNEY, AUSTRALIA

Holman House sits on the top of a 70-meter (230-foot) cliff in Sydney's exclusive Dover Height region. The kitchen and living room's curved forms, supported on four angled stilts, were inspired by the torso in Pablo Picasso's painting *The Bather*. Its lower-level stone walls form an extension of the cliff and open to terraces and a swimming pool.

House Sixel Miller

SYDNEY, AUSTRALIA

Located in the heritage conservation area of Watson Bay, this house is a nod to traditional seaside clapboard cottages with street-side verandas and simple connections to the garden. The double-height glazed box at the rear scoops in light and overlooks the garden. Other features include a timber staircase, custom kitchen, and stone tile floors. Overall, the composition is cohesive and the attention to detail is first rate.

Hrdlicka House

SYDNEY, AUSTRALIA

Designed by architects Neil Durbach, Camilla Block, and Lisa Le Van, Hrdlicka House has magnificent views of the city's iconic Harbour Bridge. Its compact layers unfold horizontally along the cliff, and glazed "piles" support pools and patios.

GLR Arquitectos

Garza García, Nuevo León, México

GLR Arquitectos is owned by Harvard University graduate Gilberto L. Rodríguez. He has been recognized in more than thirty states for his contribution to innovative building design, including commercial projects and many contemporary residential homes. Gilberto is also a professor of projects at the Monterrey Institute of Technology and Higher Education.

Casa BC

MONTERREY, NUEVO LEÓN, MÉXICO

Casa BC is a 15,608-square-foot (1,450-square-meter) home in picturesque Monterrey, and its elevated position provides far-reaching views of the National Park of Chipinque. Its distinctive geometric, cantilevered design lends a lightness to the heavy volumes defined by black granite, concrete, and exposed steel. Inside, the bedrooms, decks, home theater, and home office revolve around a double-height living room with dramatic skylights. Oriented to the sun's trajectory, the home is also a study in resource efficiency with water harvesting and graywater systems, solar water heaters, and native vegetation.

Casa CG

MONTERREY, NUEVO LEÓN, MÉXICO

Casa CG occupies a corner lot with glorious views of a park. Due to height restrictions, a basement area was excavated to meet the client's program requirements. The result is a three-level dwelling with three bedrooms and two spacious living areas. The interior materials are predominantly stone, timber, and glass.

Casa CH

MONTERREY, NUEVO LEÓN, MÉXICO

Casa CH is on a steep hill with unrivaled views of the city. The home's three levels offer the owners 9,687 square feet of premium living space including five bedrooms, a six-car garage, home theater system, wine cellar, and gymnasium. Outdoor spaces feature a covered terrace, two bathrooms, and a swimming pool.

Jackson Clements Burrows Architects

Melbourne, Australia

JCB was established by Tim Jackson, Jon Clements, and Graham Burrows in 1998, and employs about fifty people. In addition to working on large-scale global commercial projects, they have designed a variety of striking residential homes.

Applecross Residence

WESTERN AUSTRALIA

The Applecross residence, near Perth, encompasses 6,458 square feet (600 square meters) and has western views of the Swan River. The veranda (west) side of the house is defined by a double-height living space with operable louvers that control ventilation and shading, while the east side's masonry walls contain the sleeping spaces and provide thermal mass for passive heating and cooling. Interior materials consist of timber, stone, and glass.

Edward River House

NEW SOUTH WALES, AUSTRALIA

Edward River House is on the Edward River near Deniliquin. The design is a contemporary twist on the traditional western red cedar-clad farmhouse, with contrasting materials such as concrete, corrugated steel, and glass adding an industrial appearance. Utilitarian areas such as the garage, laundry, and mud room are on the ground level; living spaces are upstairs, where wood flooring and cabinetry complements the surrounding landscape.

Kew House

VICTORIA, AUSTRALIA

Kew House is in the picturesque suburb of Kew and overlooks the Yarra River and Studley Park Golf. Suspended on columns in a tree canopy, its steel and copper titanium alloy–clad form telescopes from front to back and contains an entry shell with study and garage, a middle shell with bedrooms, bathrooms, and laundry, and an end shell with kitchen, dining, and living areas. This dwelling sympathetically combines a variety of materials such as timber, steel, oxidized zinc, and glass.

Kobi Karp Architecture

Miami, Florida

Kobi Karp is a world-renowned firm that has been designing exquisite homes for more than two decades. With a second office in the Middle East, they service an extensive list of clients worldwide.

Circle Residence

MIAMI BEACH, FLORIDA

Situated on the Biscayne Bay waterfront, this majestic home draws inspiration from some of Florida's finest modern buildings and tropical architecture from the past. It is designed to maximize cross-ventilation, eliminating the need for air conditioning.

Day Residence
MIAMI, FLORIDA

This 10,000-square-foot home contains four master bedroom suites, two kitchens, terraces, exterior living spaces, a family room, guest suite, spa, and pool. Outside are stunning landscaped gardens and a basketball court.

Matheson House

MIAMI, FLORIDA

Matheson House is a series of floating, glassy rectangles on a prime waterfront location overlooking South Miami and the natural mangroves of Biscayne Bay. It is a modern interpretation of a traditional Florida-tropical home on wooden stilts, with verandas, a rooftop terrace, and a bayside hammock area in the shade under the house.

Laidlaw Schultz Architects

Corona del Mar, California

Laidlaw Schultz Architects was founded in 1992 by Scott Laidlaw and Craig Schultz. The company employs a close-knit team of associates that works on both commercial and residential buildings. Notable commercial achievements include Cedars-Sinai Medical Center, Paramount Pictures, 20th Century Fox Studios, and Warner Bros. Studios.

Brown Residence

NEWPORT BAY, CALIFORNIA

A simple palette of board-formed concrete, Texas shellstone, ipe siding, and white plaster helps this house blend into its natural environment overlooking Newport Bay. The property includes a large cavern with wine storage and tasting rooms to indulge the client's wine-making hobby.

Christie Peck

LONG BEACH PENINSULA, CALIFORNIA

This beachfront home for a California-born, water-loving owner is raised on a concrete and stone plinth for protection from periodic rising tides and the boardwalk traffic below. Its open, double-height plan takes advantage of ocean views. The living room is the centerpiece, a primary gathering spot surrounded by an office and master bedroom.

Gwaltney

EMERALD BAY, CALIFORNIA

This archetypal California beach home is clad in cedar, recalling the lattices around many outdoor decks. The interiors subtly incorporate the client's sizable art collection. French oak flooring, wood kitchen cabinetry, custom zinc countertops, and exposed concrete walls lend the natural, earthy feel of a true beach house.

Nico van der Meulen Architects

Johannesburg, South Africa

Nico van der Meulen is an innovative architectural practice founded in 1984 by married couple Nico and Santa van der Meulen. Over the last three decades, the business has grown substantially and now employs a team of top-notch architects and interior designers. The firm's primary focus is on cutting-edge design and functionality.

House in Blair Atholl

JOHANNESBURG, SOUTH AFRICA

House in Blair Atholl is a modern interpretation of the classic farmhouse; its floating, pitched slate form and the use of concrete, glass, and steel give the home a contemporary twist. In addition to its 750 square meters (8,000 square feet) of living space, the home's spacious lanai with a bar and seating and adjacent swimming pool create the ideal setup for outdoor entertaining.

Glass House

JOHANNESBURG, SOUTH AFRICA

This home's innovative design—a horseshoe shape around a central water feature—helps to mitigate its grand scale of 2,068 square meters (22,260 square feet). The architect used a broad palette of organic materials like stone and wood to complement the expanses of glass and concrete. Interior highlights include a gymnasium, billiards room, home theater, garage for fourteen cars, and staff cottage. Its glamorous interior was inspired by boutique hotels around the world.

House Ber

JOHANNESBURG, SOUTH AFRICA

House Ber consists of several large cantilevered boxes enclosed by an abstract latticework of black steel bars. Glass walls and frameless folding glass doors blur the boundaries between exposure and enclosure, and interior water features add an air of tranquility. Materials such as limestone, polished marble, and Caesarstone contribute to the polished ambience.

Saunders Architecture

Bergen, Norway

Canadian-born architect Todd Saunders founded Saunders Architecture in 1998. The firm has worked on cultural and residential projects in Norway, England, Denmark, Italy, Sweden, and Canada.

Villa M

VÅGADALEN IN OS, NORWAY

Villa M perches on a cliff overlooking the water. The 500-square-meter (5,382-square-foot) home has a wraparound deck, allowing the owners to enjoy the view sheltered from the wind. In addition to a master suite and six basement bedrooms for the grandchildren, the home has a substantial, 100-square-meter (1,076-square-foot) library.

Villa Mohn

BERGEN, NORWAY

Clad in black-painted wood and natural stone, this 280-square-meter (3,014-square-foot) dwelling overlooks a large fjord. The open-plan living spaces are lifted above four ground-level bedrooms and a two-car garage to take advantage of the view, and the dining room cantilevers over the sloping site. A terrace off the living spaces is partially covered by an upper-floor library and separate terrace.

Villa Tyssøy
BERGEN, NORWAY

Villa Tyssøy is on a secluded island overlooking the sea. Built in 2014 and designed by Todd Saunders, Ken Beheim-Schwarzbach, Matt McClurg, and Rubén Sáez López, the house's footprint measures only 260 square meters (2,798 square feet), but makes a bold statement. Its innovative, softly contoured façade consists of white vertical paneling. Large windows take in the panoramic views.

Surber Barber Choate + Hertlein Architects

Atlanta, Georgia

Surber, Barber, Choate & Hertlein Architects, founded in 1971, is led by principals James Choate and Dennis Hertlein. Most of their work, focusing on preservation, restoration, and adaptive reuse, takes place in North America. However, occasionally commissions take them farther afield. Many of their buildings explore the fusion of historic and highly modern themes.

Atlanta Modern I

ATLANTA, GEORGIA

Designed to accomodate a family of seven, this formidable property offers 10,000 square feet of luxurious living and is designed around a pair of parallel stone walls. Its exterior is clad in a variety of materials including weathered fieldstone, stained cedar, cut limestone, reclaimed oak timbers, unpainted stucco, zinc cladding, and glass. The interior palette consists of wide-plank flooring, stucco, and a continuation of the stacked stone used on the outside.

Atlanta Modern II

ATLANTA, GEORGIA

Atlanta Modern II, built on a steep hill, maintains a street presence and minimizes its scale with an entry courtyard carved into a notch at the bottom of the hill, where a fountain masks the sounds of the busy street. Once inside the courtyard, a three-story stair leads to the upper levels, which overlook a pool, terraces, and a lawn with a view of the city skyline. The house's warm, natural materials include cedar, stone, unpainted stucco, and zinc panels.

Longbeach California Modern

LOS ANGELES, CALIFORNIA

This home has phenomenal waterfront views of a deep-water lagoon and the Pacific Ocean. As it occupies a tight infill site, the architects wrapped its three sides around a central swimming pool to create privacy from the neighbors and street. Public spaces circle around the pool on the main level, with bedrooms above.

UNStudio

Amsterdam, The Netherlands

UNStudio (United Network Studio) was founded by Ben van Berkel and Caroline Bos in 1988. In the last few years they expanded the business by opening offices in Shanghai and Hong Kong. In addition to residential work, the firm specializes in interior/product design, urban development, and infrastructure projects.

Collector's Loft

NEW YORK, NEW YORK

Collector's Loft is a massive Greenwich Village apartment of 5,920 square feet (550 square meters). The original concept was to design a home that could also be used as a gallery, and its minimalist design creates a tranquil environment. This has been achieved by the expansive use of natural timber, glass, and ambient lighting.

MERCURY-TIME
6 YEARS OF 116 DAYS
= 696.
JUPITER TIME
PRECESSION OF THE EQUINOXES

Le Toison d'Or

BRUSSELS, BELGIUM

Le Toison d'Or is a beautifully designed commercial and residential eight-story building with curving concrete forms and gold aluminum balconies. The largest penthouse apartment exudes colossal proportions with 750 square meters (8,073 square feet) of opulent city living. Interior features include hardwood flooring and a luxurious kitchen.

Ardmore Residence

SINGAPORE, SOUTH KOREA

Ardmore Residence is an elaborate, thirty-six-story concrete skyscraper. Its fifty-eight luxury apartments begin on the eighth level, which opens out to terraces and a swimming pool. Completed in 2013, each apartment enjoys expansive views across the city, thanks to floor-to-ceiling glazed panels, bay windows, and double-height balconies.

2. HOROLOGY

Antoine Tavan 1805

Le Locle, Switzerland

Antoine Tavan 1805 is a small, independent brand that derives its name from the legendary nineteenth-century Swiss watchmaker and is owned by Roger Zarzoso. His mission is to produce an exquisite range of handcrafted timepieces designed for the connoisseur.

Bird of Paradise Tourbillon

The Bird of Paradise Tourbillon is a beautiful example of fine Swiss craftsmanship. This watch is perfectly proportioned and has a luxurious 18-karat pink gold case measuring 46 mm. What makes this timepiece so extraordinary is the open worked dial displaying a bird of paradise piece of artwork. This fine detail is garnished with 123 diamonds (2.3 carats). Powering the watch is a mechanical hand-winding tourbillon movement with a power reserve of one hundred hours.

Eagle Tourbillon

The one-of-a-kind Eagle Tourbillon has classical proportions and an 18-karat pink gold case measuring 42 mm. Its extravagant centerpiece is a diamond encrusted (3.5 carats) eagle illustration. Other features include delicate blue hour/minute hands and an exposed tourbillon carriage. Beneath the façade is a mechanical hand-winding movement.

Horse Tourbillon

The Horse Tourbillon is more than just a watch—it is a perfectly crafted piece of jewelery. With a generous 18-karat gold case measuring 46 mm, this unique piece should appeal to the discerning collector. The central horse feature contains 223 diamonds (2.9 carats). At the heart of the watch is a sophisticated Swiss-made manual winding tourbillon movement.

Arnold & Son

La Chaux-de-Fonds, Switzerland

Arnold & Son derives its name from a legendary eighteenth-century horologist named John Arnold and his son, John Roger Arnold. The Swiss company is owned by mechanical movement producer Manufacture La Joux-Perret S. A. Within the watch industry, the brand is synonymous with the production of high-quality mechanical complications.

Constant Force Tourbillon

The Constant Force Tourbillon from the highly successful Royal Collection is a truly exceptional watch. This timepiece is well-proportioned and has an extravagant 18-karat red gold case measuring 46 mm. The open-work anthracite dial is spectacular and works harmoniously with the gold hour/minute hands. Beneath the elegant exterior lies a sophisticated thirty-nine-jewel mechanical hand-winding movement.

DSTB (Dial Side True Beat)

The DSTB (Dial Side True Beat) is a sublime watch with timeless charm. It has a luxurious 18-karat white gold case measuring 43.5 mm. Its striking façade showcases the exposed mechanical elements. Other features include a white lacquered sub dial with black Roman numerals and delicate blue steel hour/minute hands. Powering the watch is a thirty-two-jewel self-winding movement.

Time Pyramid Guilloché

Time Pyramid Guilloché is a sensational timepiece based on a historic skeleton clock (by John and Roger Arnold) dating to the early nineteenth century (1830–1845). With an opulent 18-karat rose gold case measuring 44.6 mm, the watch exudes classical proportions. What makes this particular model so special is the intricate, multi-layered guilloché pattern dial, which exposes the mechanical brilliance of the piece. At the heart of the watch is a handcrafted twenty-seven-jewel double-barrel manual winding movement.

Azimuth

Singapore, South Korea

Azimuth is an innovative watchmaker in the heart of Singapore, though its watches are made in Neuchâtel, Switzerland. The company is the brainchild of Alvin Lye and Christopher Long. Their primary objective is to create a range of distinctive contemporary timepieces for the discerning consumer.

SP-1 Carbon Crazy Rider

The Carbon Crazy Rider, released in 2016, measures a sizeable 55 mm × 38 mm × 16 mm and is designed to make a bold statement. The timepiece is partially constructed from forged carbon, making it fairly lightweight to wear. What makes this watch so impressive is the miniature chain that revolves around the center of the dial. Beneath the robust exterior lies a modified Swiss mechanical self-winding movement.

AZIMUTH
fabriqué en suisse

Mr. Roboto

After the enormous success of Mr. Roboto, the company released an updated version. This 2016 model is slightly larger than the original and has a 316L stainless steel case measuring 55 mm × 47 mm. Its eclectic design is based on the iconic Golden Robot of the 1950s. Time is presented in an unconventional retrograde format. The watch is powered with an automatic mechanical movement.

Twin Turbo

The Twin Turbo is a robust, masculine watch designed for the motoring enthusiast. With a stainless steel and aluminum chassis measuring 54 mm × 50 mm, the watch is massively proportioned. What makes this timepiece so appealing is the striking dual time display. This model also has an innovative crown cover that replicates bonnets of 1960s automobiles. At its heart are two vintage caliber ETA 2512-2 mechanical self-winding movements.

AZIMUTH

Cabestan

Presinge, Switzerland

Cabestan is an innovative watchmaker near Geneva. Its watches are renowned for their innovative designs and complicated mechanical movements. With Lionel Betoux at the helm, the company is manufacturing some of the most exclusive timepieces on the market.

Dragon & Snake Nera

Dragon and Snake Nera are two special editions (based on the Luna Nera) that were unveiled at Baselworld 2016. The watch measures a sizable 46.5 mm × 50.8 mm × 19.1 mm and has three-dimensional architectural dials with solid gold dragon and snake miniature sculptures. Beneath the luxurious façade lies a fusée-and-chain mechanical movement containing 949 individual components.

SWISS MADE

Trapezium

The Trapezium displays generous proportions and has an 18-karat gold case measuring 48 mm × 46 mm × 17 mm. Time is ingeniously displayed on vertical drums and is visible via four sapphire crystal windows. A sophisticated fusée-and-chain mechanical movement contains 882 separate parts.

12
1
2
3
45
50
55
05
10

Triple Axis Tourbillon

The Triple Axis Tourbillon, released in March 2015, has a substantial 18-karat gold case measuring 50.25 mm × 48.30 × 22.30 mm. It has a distinctive three-dimensional dial that allows the wearer to view the mechanism from all angles through three sapphire crystal windows). Powering the watch is a tourbillon movement made of 1,044 components.

Kerbedanz

Neuchâtel, Switzerland

Kerbedanz was founded by Tigran Kerbedanz, who has a passion for antique clocks. All of the company's timepieces combine the finest (Haute Horlogerie) traditional craftsmanship with original design.

KOÏ

The KOÏ is limited to only twelve pieces worldwide. This timepiece has a well proportioned 18-karat white gold case measuring 46 mm and a striking skeletonized dial with a koi sculpture, meticulously decorated and made out of gold. Beneath the façade lies a nineteen-jewel mechanical hand-winding tourbillon movement.

Polvs Arcticus

The Polvs Arcticus was released at Baselworld 2016. It has a 50 mm 18-karat white gold case and a stunning bezel garnished with eighty diamonds (6.72 carat). Other features include a grand feu enamel dial with solid gold astrolabe indication and exposed double tourbillon. Powering the watch is a forty-six-jewel mechanical hand-winding movement devised by Concepto and Kerbedanz.

Tree of Life

The Tree of Life is part of the ladies' watch collection. Its opulent 18-karat yellow gold case measures 41 mm, and a bezel encrusted with 540 emeralds makes this piece really sparkle. Other fine details include a three-dimensional glazed enamel dial with a solid gold engraved tree sculpture. The thirty-five-jewel mechanical self-winding movement is manufactured by Technotime and Kerbedanz.

Konstantin Chaykin

Moscow, Russia

Konstantin Chaykin is probably the world's most famous Russian watchmaker. In 2010, he became a member of the highly prestigious AHCI (Académie Horlogère des Créateurs Indépendants). His timepieces are manufactured in very limited editions or made to order.

Cinema

This striking timepiece pays homage to the legendary American photographer and inventor Eadweard Muybridge. With a rectangular platinum or 18-karat gold case measuring 37 mm × 47 mm × 12.2 mm, this watch is designed to make a luxurious statement. Its centerpiece is a galloping horse animation visible via a round window located at six o'clock. Other features include a black lacquered dial and open-work silver hour/minute hands. A thirty-one-jewel mechanical movement powers the watch.

Genius Temporis

This traditional timepiece has a classical 18-karat rose gold case measuring 44 mm and an exquisite hand-engraved dial. This charming characteristic contains a range of material such as brass, bronze, and steel. Other features include delicate hour/minute hands and bold black Roman numerals. Beneath the luxurious façade lies a twenty-nine-jewel mechanical movement with a power reserve of forty-eight hours.

RUSSIA

KONSTANTIN CHAYKIN

Lunokhod Prime

This timepiece has a huge 18-karat gold case measuring 50 mm. What makes this watch so unique is the large central moon phase indication constructed from a 12-mm-deep sea pearl and solid silver. Other features include a skeletonized dial with retrograde minute counter. At the heart of the watch is an in-house mechanical movement containing 346 components.

Louis Moinet

Saint-Blaise, Switzerland

Louis Moinet was an eighteenth-century horologist, painter, sculptor, and inventor of the chronograph. The modern-day company was established in 2004 by creative powerhouse Jean-Marie Schaller. The company prides itself on manufacturing a range of exquisite timepieces for the connoisseur.

Black Gold Derrick

The Black Gold is a brilliant limited edition (twenty-eight pieces) in the highly successful Derrick series of watches. With a generous 18-karat white gold case measuring 47 mm, this timepiece is designed for the serious collector. What makes this watch so spectacular is the intricate multi-layered dial. Other fine details include an animated miniature Derrick drill and exposed tourbillon carriage. Beneath the luxurious façade lies a refined, twenty-seven-jewel mechanical hand-winding movement.

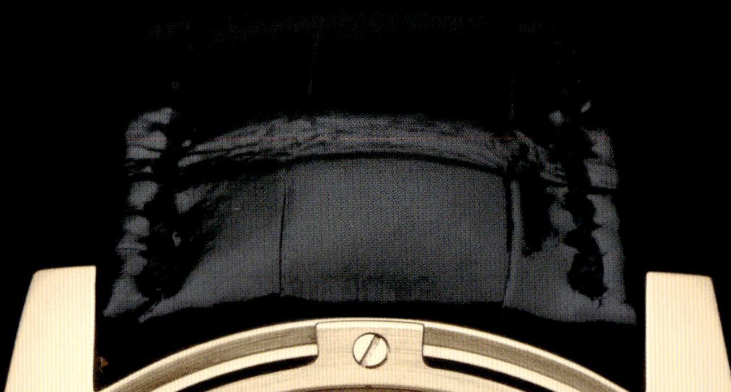

Memoris 200th Anniversary

This limited edition timepiece (twenty pieces) was released in 2016. It has an opulent 18-karat white gold case measuring 46 mm × 15.75 mm and a part skeletonized enamel dial. Other features include Gouttes de Rosée dew-drop hour/minute hands and sporty chronograph counters. Powering the watch is a thirty-four-jewel mechanical self-winding movement containing 302 parts.

LOUIS MOINET
1806

LOUIS MOINET

Sideralis

The Sideralis is a complicated timepiece synonymous with Louis Moinet. It has an extravagant 18-karat gray gold case measuring 47.5 mm and a hand-painted aventurine dial. This meticulous detail works in perfect harmony with the exposed (patented) double tourbillon carriages. At the heart of the watch is a sophisticated thirty-six-jewel mechanical hand-winding movement.

Patek Philippe

Geneva, Switzerland

Patek Philippe has a reputation as one of the world's finest watchmakers. The company has kept its independence for nearly 170 years and is privately owned by the Stern family. Frequently its timepieces achieve unbelievable prices at auction and are coveted by serious collectors. Manufactured in Switzerland, the watches undergo stringent quality tests.

5930G-001

The 5930G-001 is a spectacular creation designed for the most discerning watch collector. This timepiece displays modest proportions and has an 18-karat white gold case measuring 39.5 mm. What makes this watch so remarkable is the intricate hand-guilloched blue opaline dial. Other striking features include raised gold indexes and twenty-four-hour time zone indication. The timepiece has a mechanical self-winding movement comprising 343 components.

6002G-010

Measuring 44 mm, this watch is designed to make a statement without being overbearing. It features a meticulously engraved 18-karat white gold case and a decorated black enamel (champlevé/cloisonné) dial with raised gold Roman numerals and moon phase indication. Powering the watch is a mechanical hand-winding movement containing 705 parts.

6300G-001

This wonderful watch, from the highly successful Grand Complication range, has an 18-karat white gold case measuring 47.4 mm and an ebony black opaline dial with hand-guilloched hobnail-patterned center. The massive list of functions includes a minute repeater, leap year cycle, moon phase, and second time zone indications. Beneath the polished exterior is a manual-winding mechanical movement containing 1,366 parts.

PATEK
PHILIPPE
MOUVEMENT
SONNERIE
LUNE
Date

PATEK PHILIPPE
MOUVEMENT
SONNERIE
GENÈVE
SWISS MADE

Rebellion Timepieces are part of an exciting new breed of high-end contemporary watchmakers. Situated on the banks of Lake Geneva, the company has dedicated itself to designing and manufacturing watches for the absolute perfectionist.

Gotham

The Gotham has epic proportions, measuring 52.2 mm × 47.9 mm × 18.2 mm. It has an almost robotic looking titanium case. Other features include twin barrel cylindrical hour/minute indications and an integrated manual winding lever. Powering the watch is a mechanical movement with a staggering power reserve of 1,000 hours.

Magnum 540

This watch has a massive forged carbon case measuring 46 mm × 56.7 mm × 19.6 mm. A complicated, multi-layered dial reveals its mechanical workings—a thirty-eight-jewel tourbillon movement containing 490 components.

T-1000 2

The T-1000 2 is a brilliant timepiece that should appeal to connoisseurs of modern design. With a lightweight grade 5 titanium case measuring 46.7 mm × 46.9 mm × 19.5 mm this watch will surely attract attention. What makes this timepiece so special is the innovative design and attention to detail. Other details include a marvelous architectural three-dimensional dial with twin barrel cylindrical hour/minute display. At the heart of the watch is an accomplished 39-jewel mechanical movement comprising 693 separate parts.

REBELLION

RPaige Watch

Honolulu, Hawaii

RPaige is the brainchild of fourth-generation watchmaker Richard Paige. Prior to establishing the company, Paige owned several watch stores in California and founded the groundbreaking website TimeZone.com. He has also designed art deco–inspired furniture, jewelry, glassware, and art, which is sold in many exclusive retail outlets throughout the United States.

"Crash of '29"

The "Crash of '29" limited edition includes forty-nine pieces worldwide and was designed by Richard Paige and Mark Carson. Its art deco-inspired polished steel case works in perfect harmony with the sleek black dial. Other features include bold white numerals, small seconds indication (located at six o'clock), and a top-mounted crown. It also contains a vintage American pocket watch mechanical hand-winding movement.

"DuoFace" Elgin from 1903

The DuoFace collection is a series of timepieces with distinctive vintage dials. This model has a beautifully crafted antique white enamel dial produced by American manufacturer Elgin. It has a unique articulated, hinged polished steel 44 mm case that allows the timepiece to be worn face up or down. Other features include two sapphire crystal lenses and an ornate decorative crown. Beneath the stunning façade is a vintage mechanical hand-winding movement produced by Elgin in 1903.

ELGIN

Wrocket Skeleton
Antique Elgin Movement

The Wrocket series was RPaige's first collection and epitomizes the philosophy behind the brand. This superb creation is one of the most ornate and beautiful versions to date. It has a stunning antique skeleton dial hand-engraved by the finest European craftsman. Other features include trademark art deco–inspired lugs, blued steel hands with red enamel tips, and Roman numeral markers. Powering the watch is a fifteen-jewel Elgin antique American pocket watch dating to 1905.

Vacheron Constantin

Geneva, Switzerland

Vacheron Constantin is synonymous with manufacturing highly prestigious watches for the discerning collector. The business was originally established in 1755 by independent watchmaker Jean-Marc Vacheron and is now owned by the Richemont Group. Currently the company employs roughly 1,100 people, who are dedicated to producing superlative Swiss-made timepieces.

Maitre Cabinotier Retrograde

Released in 2016, this watch has a luxurious 18-karat white gold case measuring 45.5 mm × 20.06 mm. It has a stunning skeletonized dial, raised gold Roman numerals, and an exposed tourbillon carriage encapsulated in dome sapphire crystal. Powering the watch is a forty-five-jewel mechanical hand-winding movement.

ARMILLARY
GENEVE

GENEVE

Metiers D'art Fabuleux Ornements Indian Manuscript

This diminutive timepiece has an opulent 18-karat white gold case measuring 37 mm × 8.24 mm and a diamond encrusted (eighty-five gems) bezel. Other features include a hand-decorated floral dial and delicate open-work hour/minute hands. Its 18-jewel mechanical hand-winding movement is certified with the hallmark of Geneva.

VACHERON CONSTANTIN
GENEVE

Reference 57260

The Reference 57260 has been described as the most complicated timepiece ever made and is the work of three Vacheron Constantin watchmakers who dedicated eight years of their lives to this project. The result is an incredibly large (98 mm × 50.55 mm) 18-karat white gold pocket watch like no other. This piece has fifty-seven complications including a minute repeater, moon phase indicator, and Gregorian perpetual calendar. Its mechanicals are made of 288 jewels and 2,800 components.

VACHERON ✠ CONSTANTIN
5775
יום כיפור
תמוז
SIL
NIGHT
CHI

Urwerk

Geneva and Zurich, Switzerland

In the last nineteen years, Urwerk has established itself as one of the most innovative brands in the world of horology. The company is the vision of master watchmaker Felix Baumgartner and chief designer Martin Frei. Their mission is to create a range of highly complicated mechanical watches for discerning collectors.

UR-106 Black Pink

One of Urwerk's latest models, the UR-106 Black Pink targets the feminine market. It measures 35 mm × 49.4 mm × 14.45 mm and has a black (PVD treated) steel case garnished with black diamonds. Other striking details include bold pink numerals and a stunning lapis lazuli moon phase indication. At the heart of the watch is a refined fifty-two-jewel mechanical automatic movement.

UR-110 Eastwood Macassar Ebony

The Eastwood Macassar Ebony is an exciting variation of the highly successful UR-110. Dimensionally the watch exudes mammoth proportions and has a black (PVD treated) case measuring 47 mm × 51 mm × 16 mm. It has a distinctive bezel with macassar ebony inlay and a revolving satellite complication with wandering hour module display. Powering the watch is a mechanical twin turbine movement comprising forty-six jewels.

URWERK
UR-110
OIL CHANGE
DAY / NIGHT

URWERK
UR-1001
(rsc)
revolving
satellite
calendar
power
days
MAR
APR
MAY
FEB

CONTROL BOARD UR-1001
1000 YEARS
1000
900
800
700
600
500
400
300
200
100
0
OIL CHANGE
100 YEARS

3. TRANSPORTATION

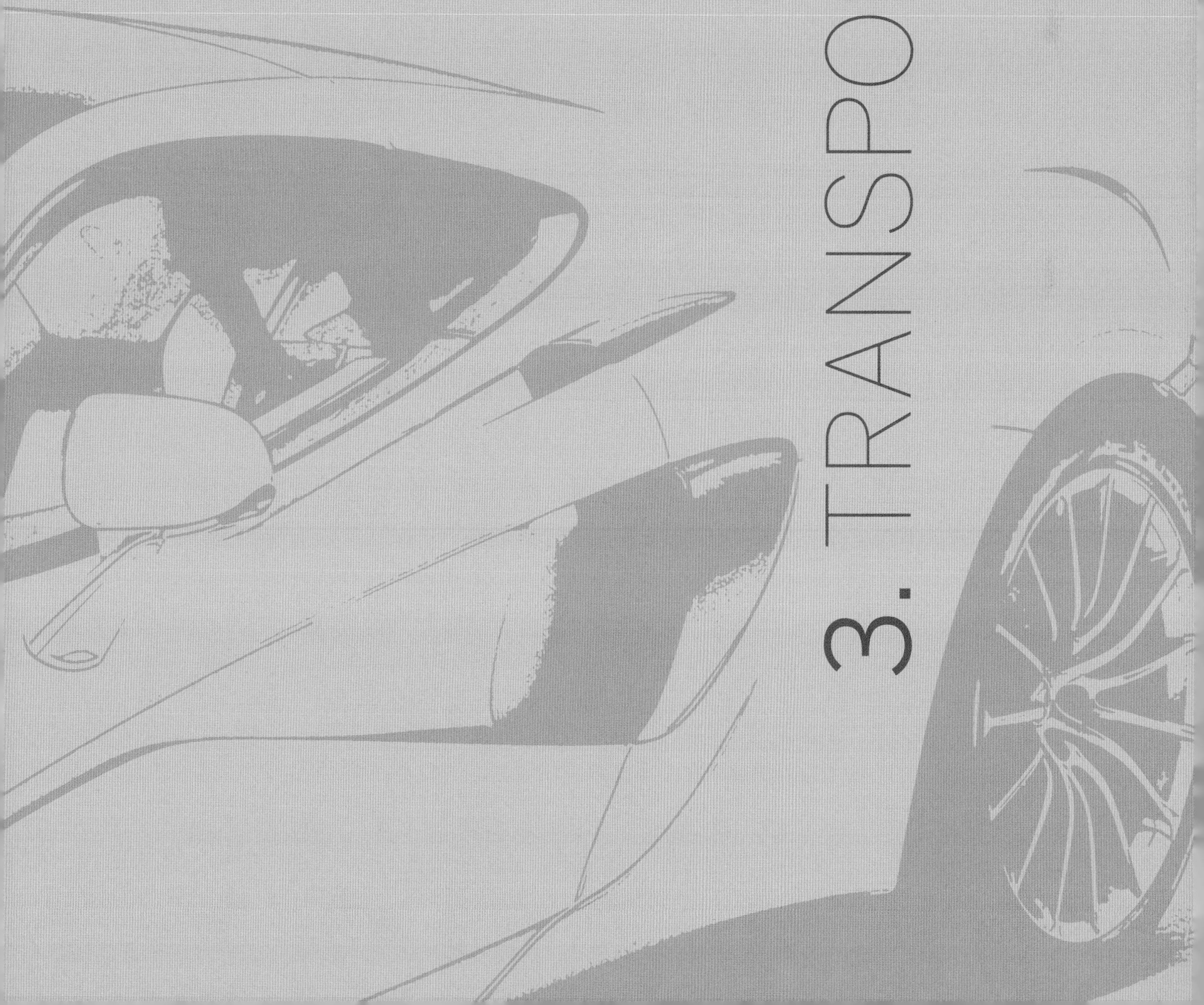

Aston Martin

Gaydon, Warwickshire, England

Aston Martin is a world-famous British manufacturer that produces a range of aspirational luxury sports cars. The business was established in 1913 by Lionel Martin and Robert Bamford but is now mainly investor-owned. Many people associate the company with its famous collaboration with the James Bond movie franchise.

DB11

The DB11 is the latest and most powerful car in the DB range. This car can go from 0–62 mph (0–100 km/h) in 3.9 seconds and reach a top speed of 200 mph (322 km/h). Primarily this is because the automobile has an impressive 5.2-liter twin turbocharged V12 engine. In addition to its stunning exterior, the vehicle has a luxurious interior featuring two-tone leather seats, forged carbon panels, and an option to upgrade to a state-of-the-art music BeoSound system by Bang & Olufsen.

Rapide S

It wouldn't be an overstatement to describe the Rapide S as one of the world's most beautiful four-door sports cars. Incredibly, this vehicle weighs only 1990 kg and has a 0–62 mph (0–100 km/h) of 4.2 seconds, due to the innovative AM29 V12 engine that delivers 552 bhp. According to the company, this is an increase of eighteen percent over the first-generation models. Interior features include sumptuous leather upholstery, integrated audio, and hands-free telephone technology.

KU15 JXX

Vanguish Carbon Edition

The Vanguish Carbon Edition is an extraordinarily good-looking car designed for the most discerning motor enthusiasts. What makes this automobile so amazing is the sleek carbon fiber chassis. On the performance front, the vehicle delivers a mean punch and powers from 0–62 mph (0–100 km/h) in 3.6 seconds. With the accelerator pressed flat, this sports car will hit a top speed of 201 mph. On the interior, the dark theme continues with black dials, leather, and carbon fiber detailing.

Auto Fabrica

Southend-on-Sea, England

Auto Fabrica is a British company specializing in the production of unique customized motorcycles. The business is the brainchild of forward-thinking brothers Bujar and Gazmend Muharremi. As each car takes six months to manufacture, its numbers are very limited, making these bikes exceptionally exclusive and highly coveted.

Type 6

The Type 6 is a highly distinctive motorcycle designed to make a bold statement. This bike is built around a Yamaha XS650 originating from 1979. However, the new version has undergone a major transformation and doesn't even resemble the original model. Notable improvements include a customized steel exhaust system, engine rebuild, and bespoke handlebars. To complete the picture, Auto Fabrica gave the bike a bare matte metal finish.

Type 8

The Type 8 motorcycle is a fully customized piece of art. It began life as a Honda CX500 dating to 1981, and has been transformed into something completely different. Auto Fabrica stripped this motorcycle back and then reconstructed it with entirely new components, including a hand-finished aluminium tank, steel exhaust pipes, and new frame.

Ferrari

Maranello, Italy

The business was established by Enzo Ferrari and is still privately owned. For nearly eighty years, the company has delighted motor enthusiasts with its extraordinarily beautiful cars. In addition to breaking the record for the world's most expensive car—the $38.1 million 1962 Ferrari 250 GTO—the company has also achieved great success in the Formula 1 world.

F12berlinetta

With its jaw-dropping good looks, the F12berlinetta is designed for the discerning car lover. It is pretty impressive on the performance front as well: with the throttle down, it will reach speeds of more than 211 mph (340 km/h). Primarily this is because Ferrari has considerably reduced the chassis weight (90KG) by using twelve types of alloy. The vehicle is upholstered in sumptuous Frau leather and has an intuitive, multi-functional steering wheel.

GTC4Lusso

The GTC4Lusso is one of the most luxurious four-seat family cars in the world. Reaching 0–62 mph (0–100 km/h) in 3.4 seconds with a top speed of 208 mph (335 km/h), it is probably also one of the fastest. This streamlined car is well-equipped with high-performance features such as a slip/slide control system, redesigned compact steering wheel, and touchscreen entertainment suite.

GTC4LUSSO

LaFerrari

The LaFerrari combines cutting-edge technology with the finest contemporary styling. This wonderful automobile will reach 0–62 mph (0–100 km/h) in less than three seconds and exceed speeds of 220 mph (355 km/h), due to its revolutionary V12 engine and advanced electric motor. In addition to being visually striking on the outside, the car has an impressive interior that includes a reconfigured digital display panel and a multi-functional Formula 1–style steering wheel.

Lamborghini

Sant'Agata Bolognese, Italy

Iconic automobile manufacturer Lamborghini epitomizes Italian flair and panache. The company, founded by Ferruccio Elio Arturo Lamborghini in 1963, was acquired by the Volkswagen Group in 1998, which is recognized as one of the world's most exciting and luxurious high-performance car makers.

Aventador SV Roadster

The Aventador SV Roadster is an exceptionally lightweight and beautifully crafted car. Its chassis is constructed from carbon fiber, monocoque, and aluminum, making it a very fast car that can accelerate from 0–62 mph (0–100 km/h) in about 2.9 seconds. The interior can be fully customized.

Centenario

The Centenario's aerodynamic exterior, finished in carbon fiber, is designed to make a sharp first impression. This vehicle is also amazingly fast and will reach a staggering top speed of 220 mph (355 km/h). The interior is meticulously hand-finished by the world's leading craftsmen.

Veneno Roadster

The Veneno Roadster is an outstanding car with a futuristic appearance. It is also highly exclusive because only nine models will ever be made. On the performance front, this car doesn't disappoint and will go from 0–62 mph (0–100 km/h) in 2.8 seconds. Like most of Lamborghini's creations, this car is constructed from lightweight materials such as carbon fiber and aluminium.

Maserati

Modena, Italy

Maserati, a luxury Italian automobile manufacturer, was launched by Alfieri Maserati in 1914, and is now owned by car giant Fiat Chrysler. Fortunately, this hasn't affected the Maserati brand, as all of the vehicles are highly designed and coveted.

GranCabrio MC

Maserati described the GranCabrio MC as the first genuine four-seater cabriolet it manufactured. Visually, this car is spectacular and is the height of luxury. Its V8 engine will hit an impressive top speed of 180 mph (289 km/h). The interior contains embossed leather seats, carbon fiber trim, and an advanced entertainment system.

Granturismo Sport

The GranTurismo Sport is a stunning coupe, and its bold design is easily recognizable as one of Maserati's creations. Powered by a V8 (4691 CC) engine, this car will reach 0–62 mph (0–100 km/h) in 4.8 seconds with a top speed of 185 mph (298 km/h). The interior is equally impressive with its Poltrona Frau leather, perforated aluminium, and carbon fiber accents.

Quattroporte GTS

The handsome and dignified Quattroporte GTS has the performance to go with its polished good looks. With the pedal down you can expect to go from 0–62 mph (0–100 km/h) in 4.7 seconds and reach a top speed of 180 mph (307 km/h).

McLaren Automotive

Surrey, England

McLaren Automotive is a British luxury sports car manufacturer founded in 1963 by Bruce McLaren Motors. In the last few years it has dramatically increased its profits and output. In 2015, the company delivered a record 1,654 automobiles to customers around the world.

570GT

The 570GT has all the charm and beauty you would expect from McLaren. With a curvaceous chassis constructed from carbon fiber, the car is also exceptionally lightweight, which helps it achieve a top speed of 203 mph (328 km/h). The car also has many superb features such as dihedral swing doors, premium leather upholstery, and a state-of-the-art touchscreen infotainment system.

McLaren
570GT

675LT Spider

The 675LT Spider is a phenomenal-looking convertible designed for absolute perfectionists. McLaren Automotive states that this is the fastest and most lightweight open-top it has produced. Certainly the statistics verify this as the car will reach 0–62 mph (0–100 km/h) in 2.9 seconds and a top speed of 202 mph (326 km/h). As you would expect for a luxury vehicle like this, it can be customized to the client's exact specifications.

675LT

P1

The P1 is designed to please the world's most discerning motor car enthusiasts. There are several models, including one with exposed carbon. In addition to its sensational good looks, the lightweight material makes the car fast. This exquisite piece of automotive engineering will reach bone-shaking speeds of 217 mph (350 km/h).

Morgan Motor Company Ltd

Worcestershire, England

The Morgan Motor Company Ltd was founded in 1909, by H. F. S. Morgan. Still privately owned, its headquarters is in idyllic Malvern. All of its automobiles are quintessentially English and are reminiscent of bygone days. However, they are also equipped with high-performance engines and the latest automotive technology.

Aero 8

The Aero 8 has an effortless charm that will certainly appeal to motoring purists. This wonderful car has vintage styling, combined with an enormous amount of raw power. This car has a refined BMW 4799CC V8 engine that will reach speeds of 167 mph (270 km/h), making for a truly exhilarating experience. Its interior combines organic materials such as woven fabric, wood, and leather.

Plus 8

The Plus 8 has a timeless charm that will appeal to motoring connoisseurs. This car is available with a manual gearbox as standard and an optional 6-speed ZF automatic transmission if desired. Whatever the preference, its performance won't disappoint, as the BMW 4799CC V8 engine will hit a top speed of 155 mph (249 km/h). Other specifications include tanned leather seats, six spoke alloys wheels, and an easily removable roof hood.

MORGAN

Roadster

The marvelous-looking Roadster wouldn't look out of place in an old British movie. This creation has all the hallmarks of a vintage car, with an abundance of modern technological features. Specifications include chrome spoke wheels, cream leather upholstery, ash wood trim, and power steering. Beneath the bonnet lies a powerful Ford engine that will go from 0–62 mph (0–100 km/h) in an impressive 5.5 seconds.

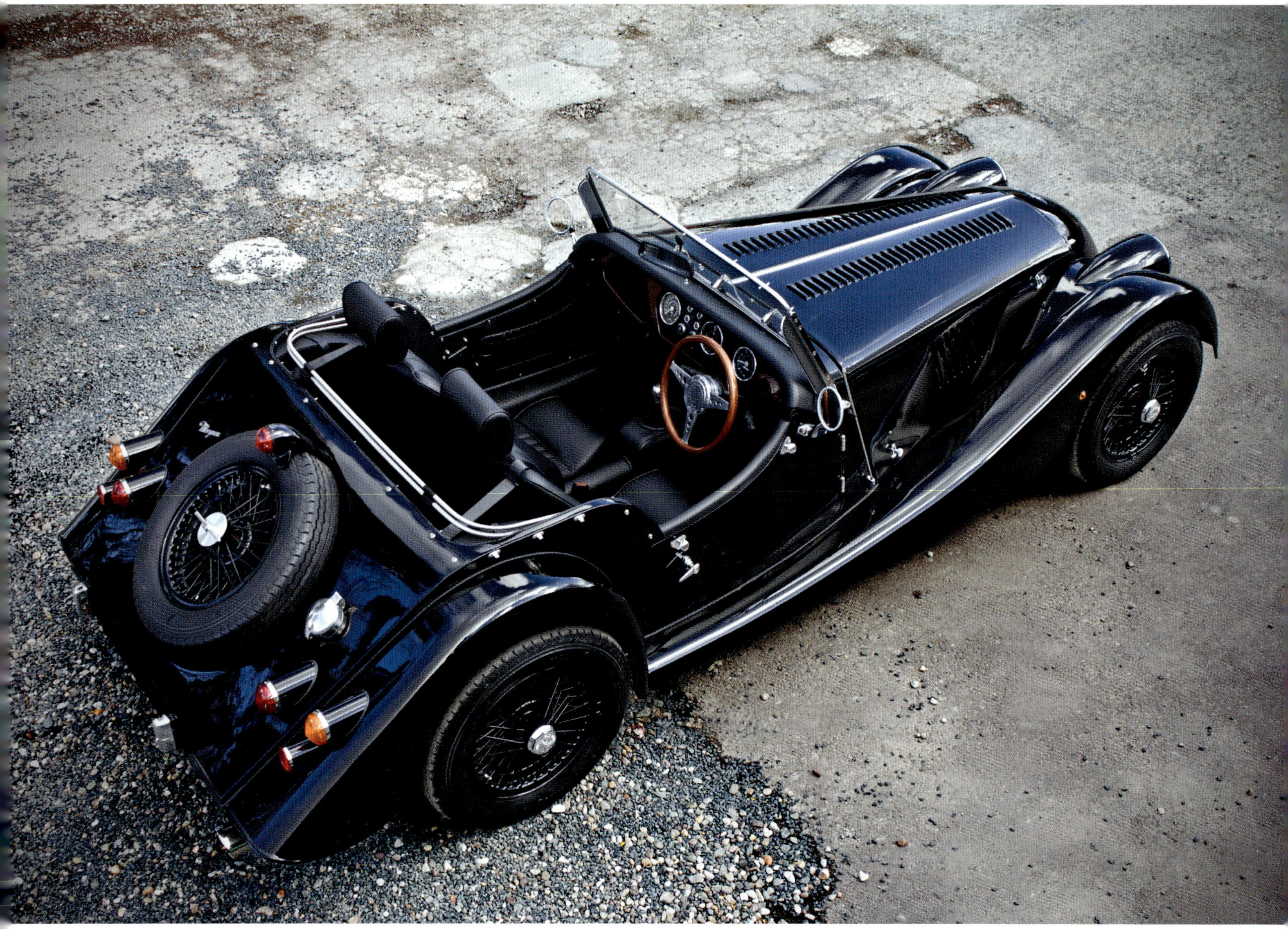

MORGAN

Pagani Automobili S.p.A.

Modena, Italy

Pagani Automobili S.p.A is the brainchild of Argentinian visionary Horacio Pagani and was established in 1992. He has used his experience working for Lamborghini to develop a range of beautifully handcrafted vehicles.

Huayra BC

The Huayra is an astonishing creation designed for the ultimate car aficionado. This car has a curvaceous chassis constructed from carbon fiber and titanium, making it exceptionally lightweight and fast. The car is equipped with a 5980cc (730 hp) Mercedes-AMG M158 V12 twin-turbo engine that will go from 0–62 mph (0–100 km/h) in 3.3 seconds and reach a top speed of 223 mph (360 km/h).

Zonda Cinque Roadster

The Zonda Cinque Roadster is a limited-edition (five units) open-top car originally unveiled in 2009. This car boasts an innovative frame constructed from high-tech materials such as carbon and titanium. The lightweight chassis and Mercedes AMG V12 engine enable the vehicle to travel from 0–62 mph (0–100 km/h) in 3.4 seconds. The interior features quality leather and carbon-fiber upholstered seats.

Zonda Tricolere

The Zonda Tricolere was released in 2010, to celebrate the 50th anniversary of the "Frecce Tricolori" and achievements of the National Aerobatic Team. This wonderful car features a sophisticated carbon-titanium monocoque chassis, titanium/magnesium suspension system, and Cima six-speed sequential gearbox. It also delivers on the performance front and will accelerate from 0–62 mph (0–100 km/h) in an impressive 3.4 seconds.

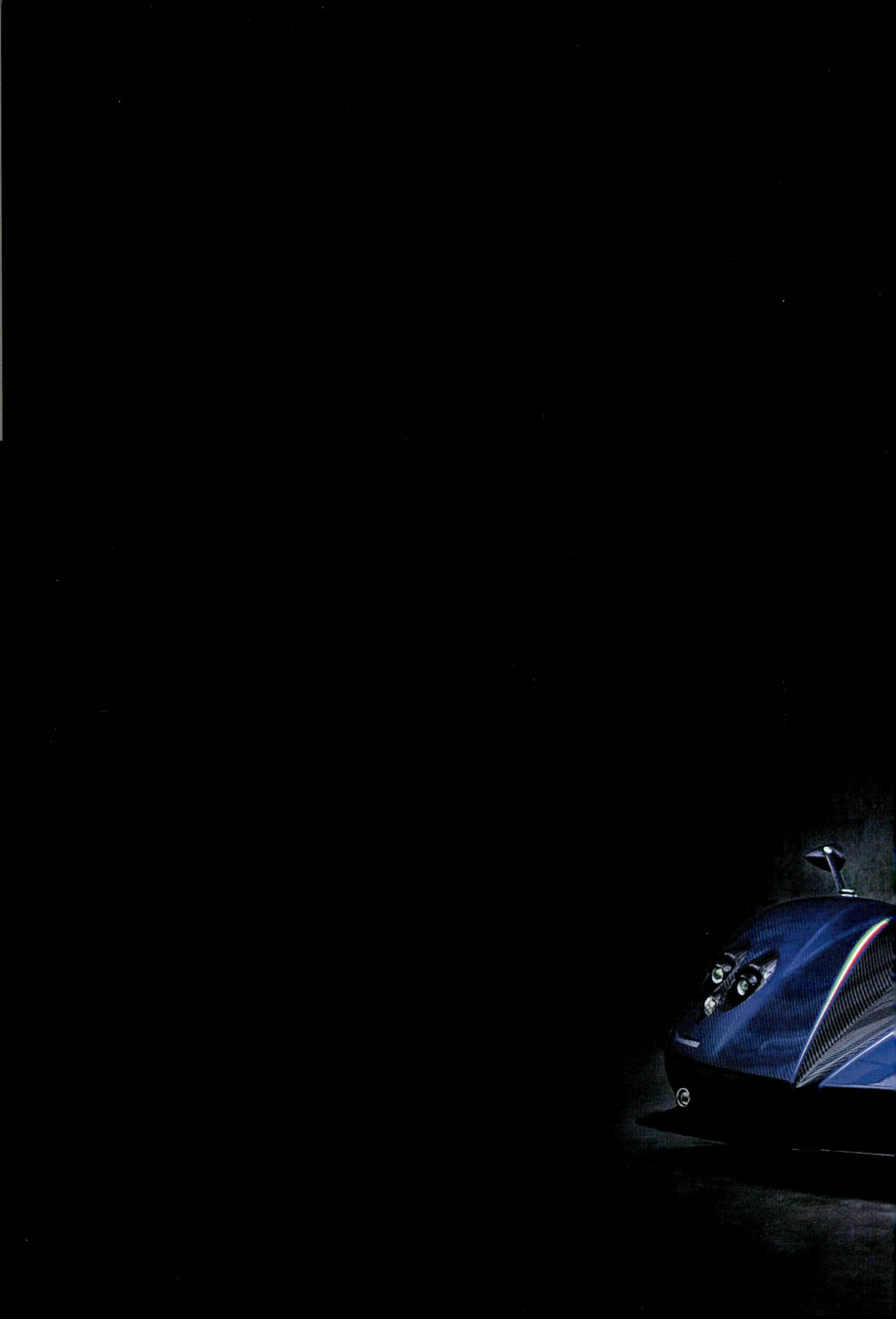

P ZERO
P ZERO

Rolls Royce

Derby, England

To many people, the name Rolls Royce symbolizes the epitome of good taste and class. Established in 1906 by Charles Stewart Rolls and Sir Frederick Henry Royce, the company is still British-based and headquartered in Derby, England. For more than a century, the company has dazzled the public with its exquisite range of luxurious automobiles.

Dawn

Dawn is a magnificent convertible conceived for the discerning car enthusiast. It is beautifully designed right down to the most meticulous detail. With four spacious seats, it is also an incredibly luxurious family car. As you would expect from Rolls Royce, the automobile has an abundance of extravagant features such as sumptuous leather seats and hardwood detailing.

Ghost Black Badge

The Ghost Black Badge is a seductive car that looks different from many other models in the range. The enhanced 6.6-liter twin-turbo engine makes it the fastest Ghost to date. Its interior has some truly inspirational features such as a carbon fiber dashboard, three-dimensional forged aluminum wheel trims, and two-tone leather seats.

Phantom Drophead Coupé

The design of Phantom Drophead Coupé is inspired by the J-class racing yachts of the 1930s and takes luxury to a whole new level. Every vehicle is fastidiously hand-built by the world's finest craftsman, and every model can be personalized to the client's exact specifications. Powering the car is a high-performance, 6.75-litre V12 engine that will reach 0–62 mph (0–100 km/h) in only 5.7 seconds.

Schöpfer Yachts

Boston, Massachusetts

Schöpfer Yachts is a luxury boat producer in pursuit of excellence. Founded in 2008, the company is the brainchild of innovative designer Kevin Schöpfer, who collaborates on each project with the world's finest naval architects. The result is an astonishing range of super-yachts designed for the elite consumer.

Audax

The Audax 130 is a beautiful sports yacht designed by Kevin Schöpfer and naval architects Sparkman & Stephens. This sublime vessel offers an unbelievable level of luxury that includes spacious lounge/dining/entertainment areas and a carbon fiber tuna tower. Overall, the futuristic appearance is cohesive and the construction is first-rate.

Infinitas

The modernistic design of the Infinitas will certainly appeal to yacht connoisseurs. It can accommodate sixteen guests, and its minimalist interiors wouldn't look out of place in a sci-fi film. Other features include a private pool, opulent master suite, and helicopter pad.

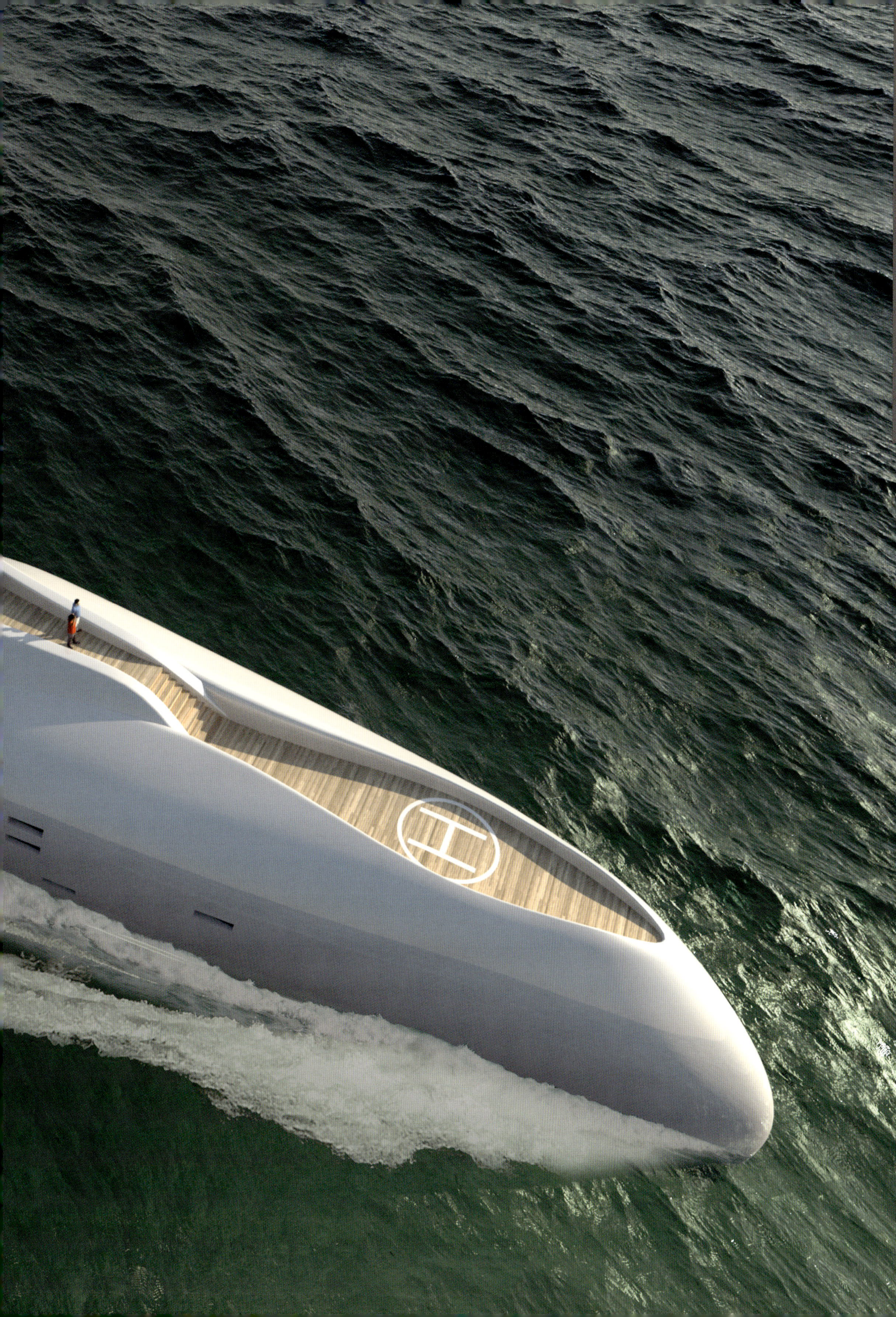

Oculus

The Oculus, Kevin Schöpfer's first yacht design, was inspired by the jaw and eye sockets of large oceanic mammals. This gives the yacht an imposing and unique appearance. Its interior offers ample living spaces, including a second-level owner's suite accessed by a central spiral staircase or elevator tube.

Thunderbike

Hamminkeln, Germany

Thunderbike is a specialist motorcycle company owned and operated by bike fanatic Andreas Bergerforth. All of its creations are distinctive and customized to the client's specifications, making each model unique.

Life Fitness

The Life Fitness is an incredible-looking dragster designed for serious motorcycle enthusiasts. This amazing bike was created in 2012, and is powered by a Harley-Davidson engine. The futuristic frame has an almost art-like appearance. Other features include bespoke brake discs, aluminum monoblock solid wheel trim, and perfectly formed exhaust pipes.

Life Fitness

Thunderbike

Radical Racer TB RR Production-R v2.0

The Radical Racer TB RR Production-R v2.0 has a bold and muscular appearance. The sculpted frame is meticulously handcrafted by the finest German artisans. Other details include three-dimensional architectural wheel trims, an ergonomic curved seat, stylized custom handlebars, and an HD 110 Screamin Eagle engine.

Unbreakable

The Unbreakable is a staggeringly beautiful bike with an abundance of masculine appeal and a distinctive art deco-inspired appearance. Detailing on the open-work trapeze forks and exposed polished engine is absolutely sensational. Every aspect of this mechanical marvel has been fastidiously hand-finished by the finest craftsmen.

Unbreakable
the MONSTER

4. LIFESTYLE

Choisi Suisse

Bienner, Switzerland

Established in 1929, Choisi Suisse has been recognized for its many achievements in the world of horology. In 2013, new owners took the company in a different direction. While retaining many traditional values, the present-day Choisi is now more of a chic lifestyle brand than a watchmaker.

1929 GT

The 1929 GT is a playful twist on the conventional military-style watch. Its design was inspired by 1920s timepieces and updated with a sleek black dial with accents of blue, orange, and white. This detailing works in harmony with the brushed steel 44 mm cushion-shaped case. Other features include open-work hour/minute hands and a small second counter (located at six o'clock). This cool timepiece could be worn with formal or casual attire.

12
9
3
6
SUISSE

1929 Military Original

The Military Original is a fine-looking timepiece from Choisi Suisse's highly successful 1929 collection. Dimensionally, the watch is well proportioned and has an elegant brushed steel case measuring 44 mm, giving it an androgynous appeal. What makes this timepiece so appealing is the stylish, two-tone black and copper dial. Other features include a small second indication, delicate hour/minute hands, and calf leather strap.

Heritage Tachygraph

The Tachygraph is one of the flagship models from the distinctive Heritage series of watches. Essentially what Choisi has created is a contemporary pilot watch, which oozes charm and sophistication. With a refined stainless steel case measuring 42 mm (excluding lugs and pushers) the watch will appeal to a wide audience. What makes this timepiece so special is the intricate white dial with Tachymeter scale. Other interesting details include two chronograph counters, discreet gold-colored hour/minute hands, and a stylish brown calf leather strap.

DARWINmachine is an exclusive brand specializing in the production of high-end computers. The company is the brainchild of architectural designer Matthew Kim. All of his products are highly innovative and unlike anything else on the market. They are manufactured in California in small numbers.

Hammerhead FMR989

The Hammerhead FMR989 computer has a futuristic, deconstructed appearance that reveals the machine's intricate electronic workings. The chassis is handcrafted from 6061 aircraft aluminum with translucent quattro carbon ecoresin panels. This computer measures 21 inches long by 17 inches high by 7.5 inches wide and can be customized.

Proteus

The Proteus is the latest and smallest offering from DARWINmachine. This computer is considerably more compact than the Hammerhead or Slipstream models. However, the hardware can be adapted to deliver enough power to satisfy the client's requirements. The modernistic frame is constructed from high-performance titanium, which makes it lightweight and portable. This computer will also be available in aircraft aluminum.

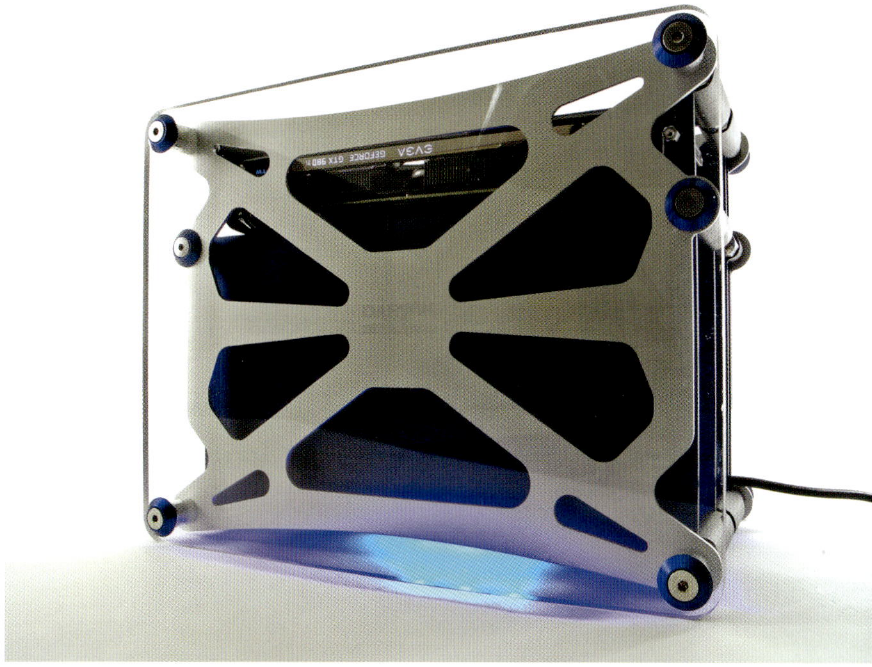

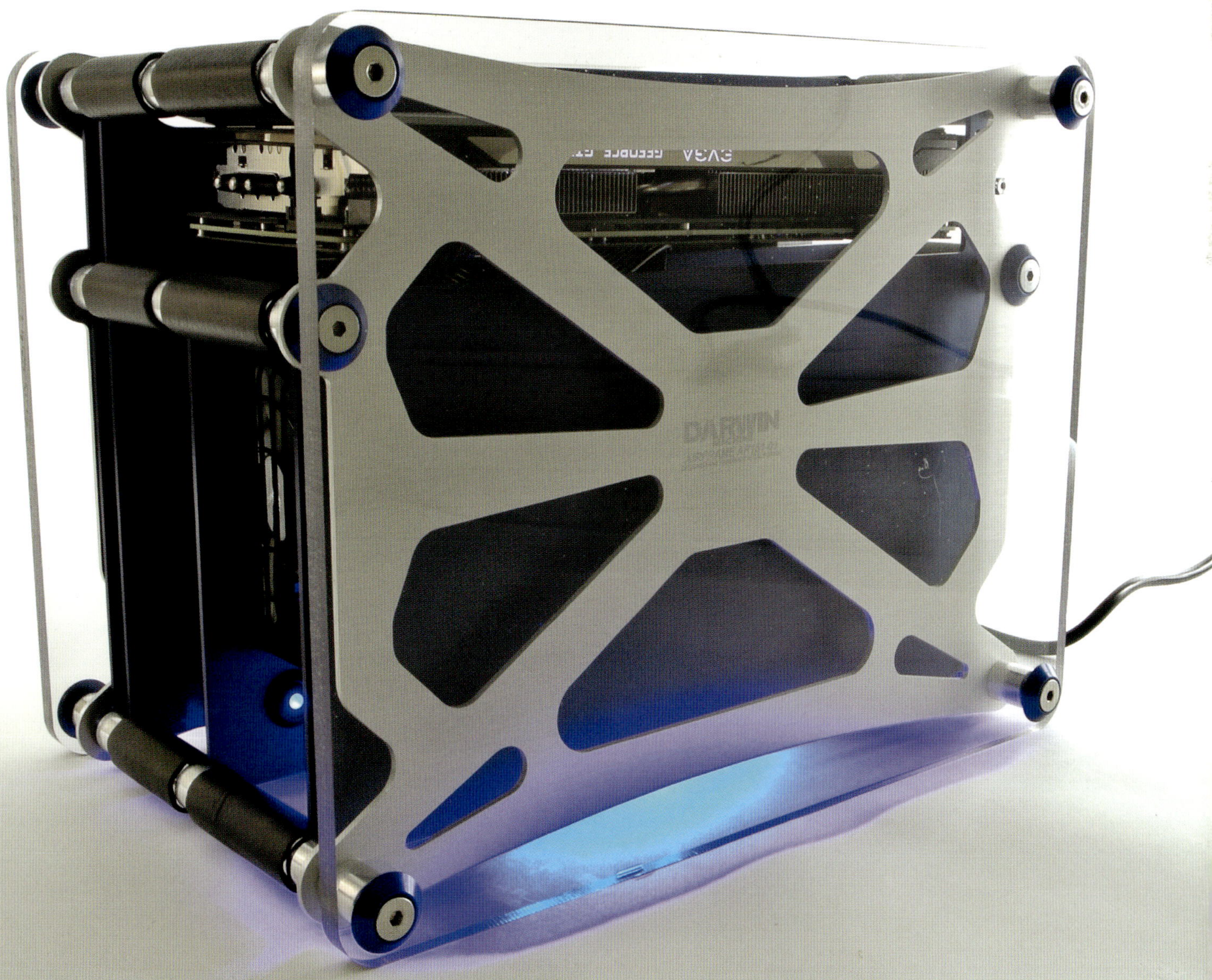

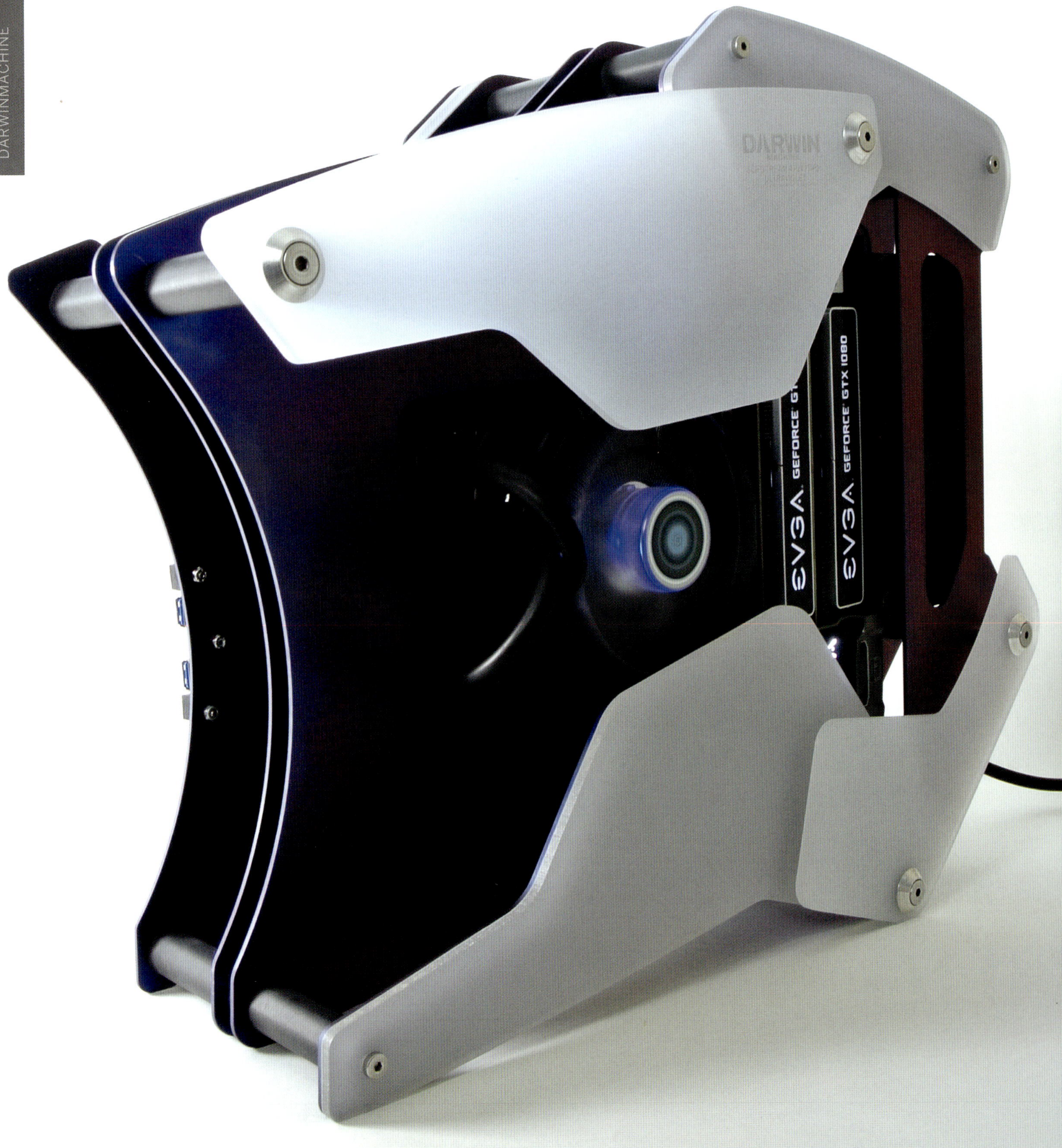
DARWIN
EVGA GEFORCE GTX 1080

AP242T Slipstream Titanium

The AP242T Slipstream is DARWINmachine's flagship model and will be limited to fifty units worldwide. Its sublime brushed, heat-treated, blued titanium gives this exclusive product a highly distinctive industrial appearance. With dimensions of 19.5 inches long by 16 inches high by 9 inches wide, this model is slightly smaller than the Hammerhead, but its hardware can be configured to meet the customer's demands.

EcoSmart Fire

Sydney, Australia

EcoSmart Fire is a high-end fireplace manufacturer just outside Sydney, Australia. Its products are exceptionally well-designed and are powered by clean, green bioethanol fuel. This means that they don't require a flue or vent pipe and can be installed just about anywhere.

Ghost

The minimalist Ghost fireplace, designed by Marc Philipp Veenendaal, is constructed of glass and appears almost invisible. Its compact size makes it ideal for even the smallest homes, and its efficient 2.5-liter burner will last up to eight hours.

Stix

With its clean, contemporary lines and brushed steel surface, Stix makes a statement in the most intimate outdoor living areas. This fabulous product won the 2011 Good Design Award (Australian International Design Awards) and was created by acclaimed Barcelona designer Hiroshi Tsunoda in collaboration with EcoSmart.

Wharf

Unveiled in 2016, Wharf is one of the latest products in the company's line of outdoor fireplaces, and is a perfect architectural finishing touch for courtyards, patios, pool decks, and terraces. It has a fuel-burning capacity of up to thirteen hours. This model combines durable industrial concrete composite and grade 304 stainless steel.

IL Hoon Roh

Seoul, South Korea

IL Hoon Roh designs exclusive furniture in Seoul. After receiving a degree in architecture, he earned a master's in product design at London's prestigious Royal College of Art.

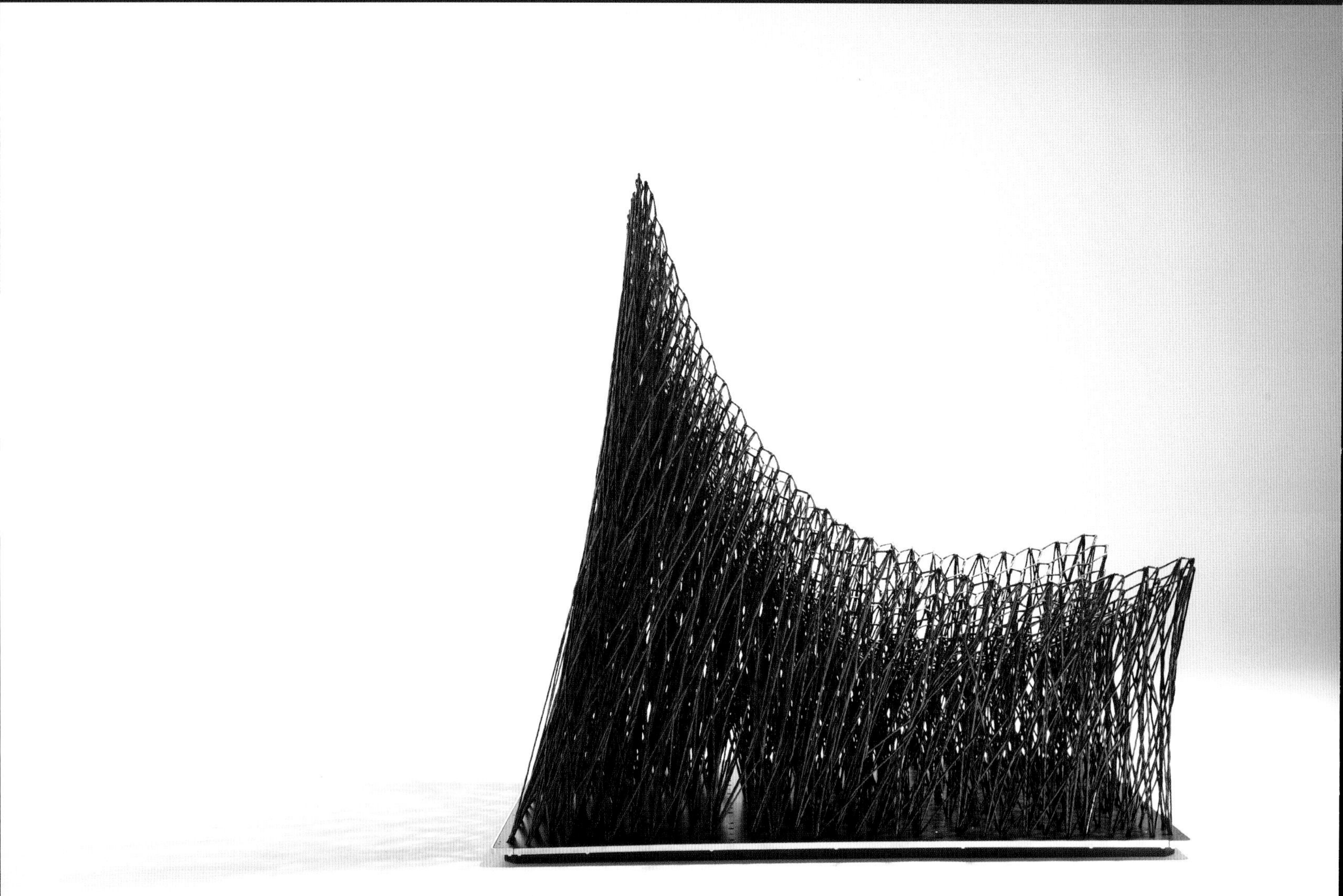

Luno

Luno is the Latin word for curve, an apt description of this limited-edition (ten pieces) armchair. Part of the innovative Rami series, it is constructed from several strands of carbon fiber, giving it an organic appearance. The artist found inspiration in the rigidity and lightness of a bird skull.

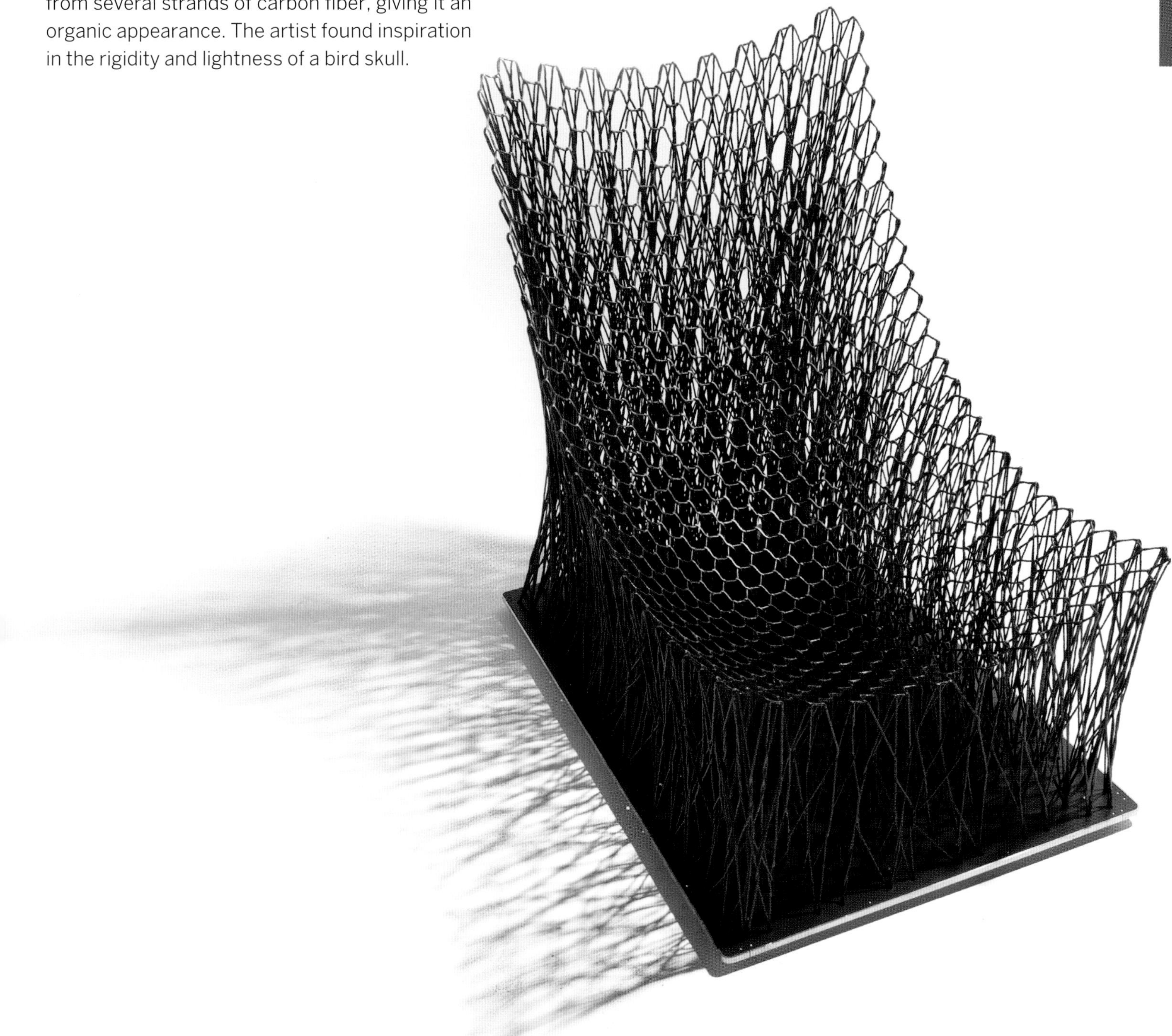

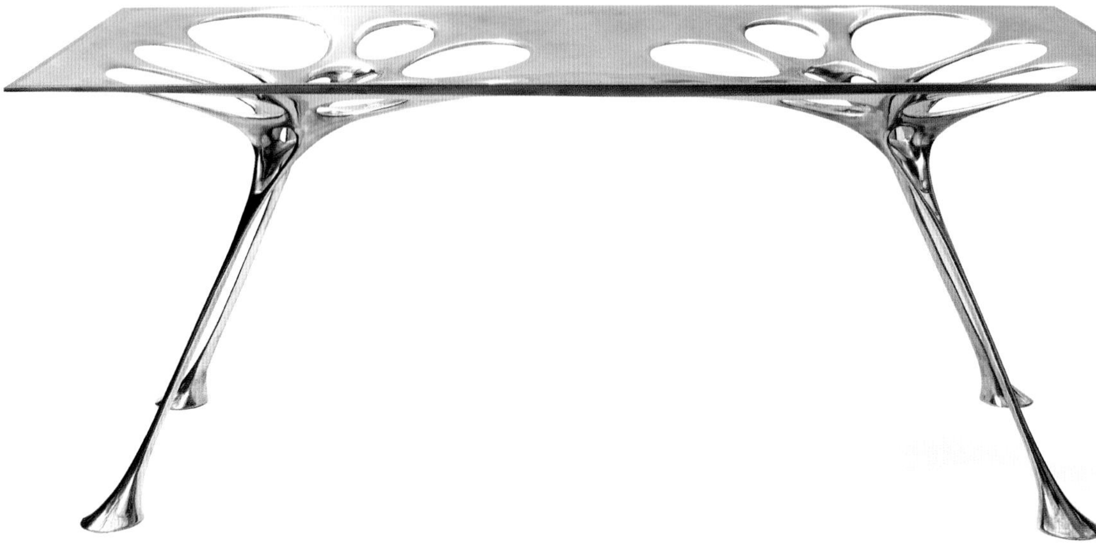

Table R EX 08

The Table R EX07 (Table Radiolaria Experiment07) has a wonderful organic appearance. It is handcrafted from aluminum that has been stretched to create a pattern of geometric holes. IL Hoon Roh collaborated with leading mathematicians to achieve this revolutionary look.

Velo

The word Velo means "covered up" in Latin and is the inspiration behind this exquisite bench, limited to only eight pieces. Hand cast from solid bronze, it creates the illusion of flowing fabric, which gives the piece a seemingly weightless appearance.

iXoost

Modena, Italy

iXoost is a digital audio equipment manufacturer whose luxury iPod docks are entirely made by hand and are unlike anything else on the market. Combining a passion for music and automobiles, it is the only company in the world fabricating iPod docking stations from supercar components. The result is a range of music players designed for a discerning clientele.

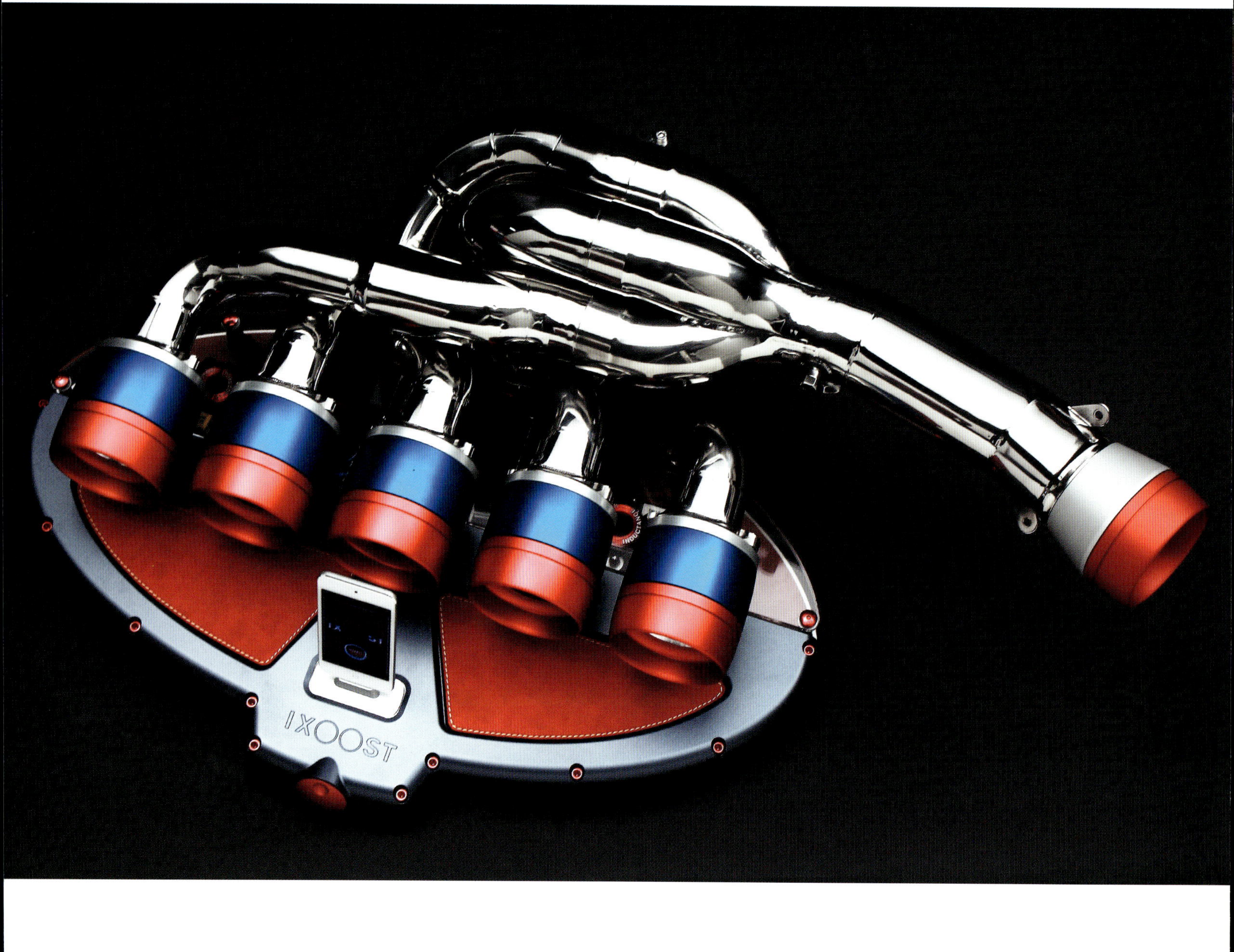

Des Jaguar

The sensational Des Jaguar is from the popular Otto range, and this iPod docking station was custom made to the client's specifications. The design pays tribute to the motorsport industry and features an innovative exhaust tail pipe as the amplifier. With an integrated speaker pumping out 240 watts, it also delivers on the performance front. Other fine details include red leather accents, a brushed aluminum base, and Bluetooth connectivity.

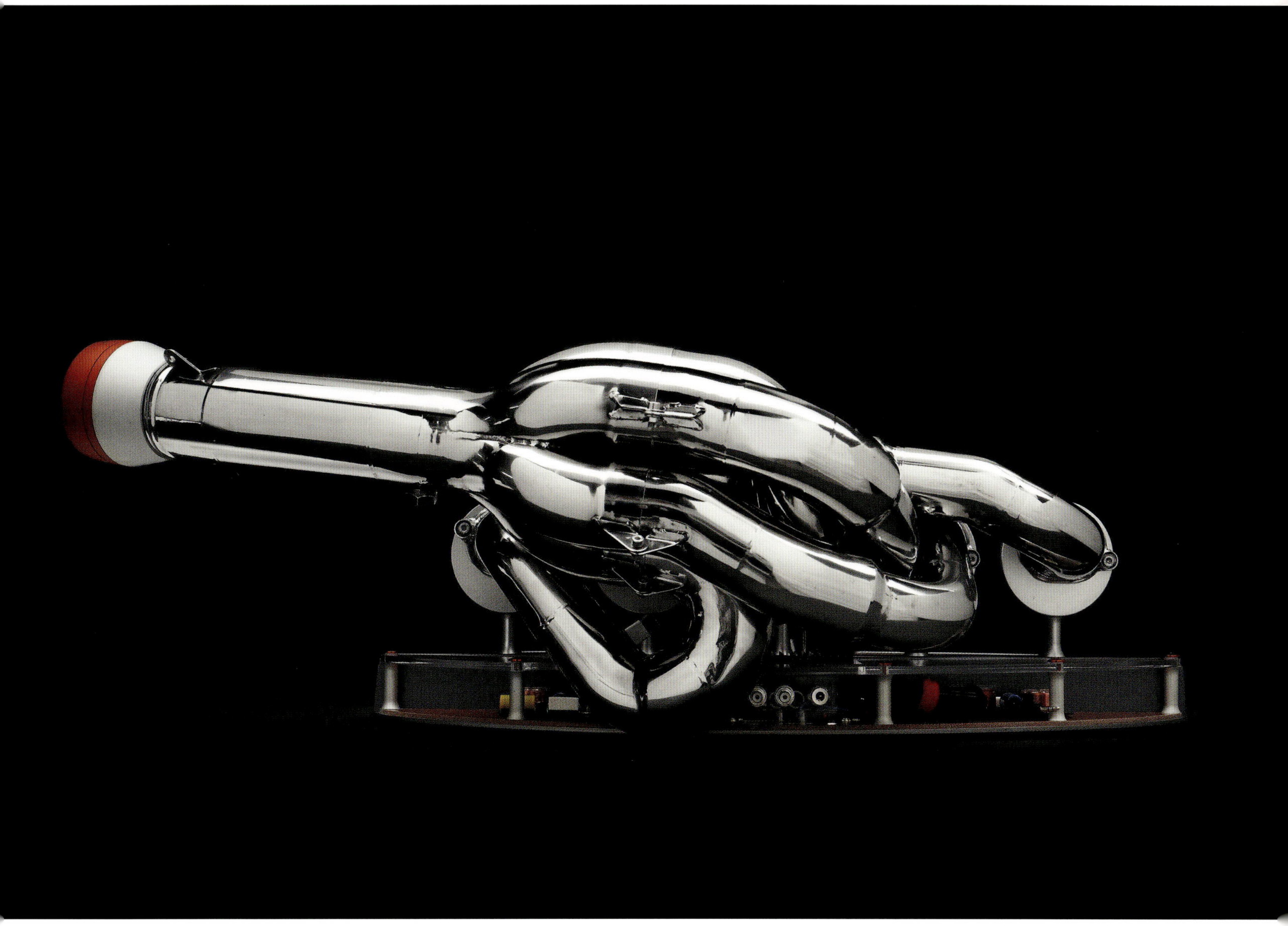

Radial6

The Radial6 is a distinctive piece of audio art that draws inspiration from the aviation industry. Like all of iXoost's creations, it can be custom made. Its striking radial speaker configuration enables the device to distribute the noise evenly, thus creating a harmonious three-dimensional sound. Other features include polished tailpipes and a phenomenal central cone.

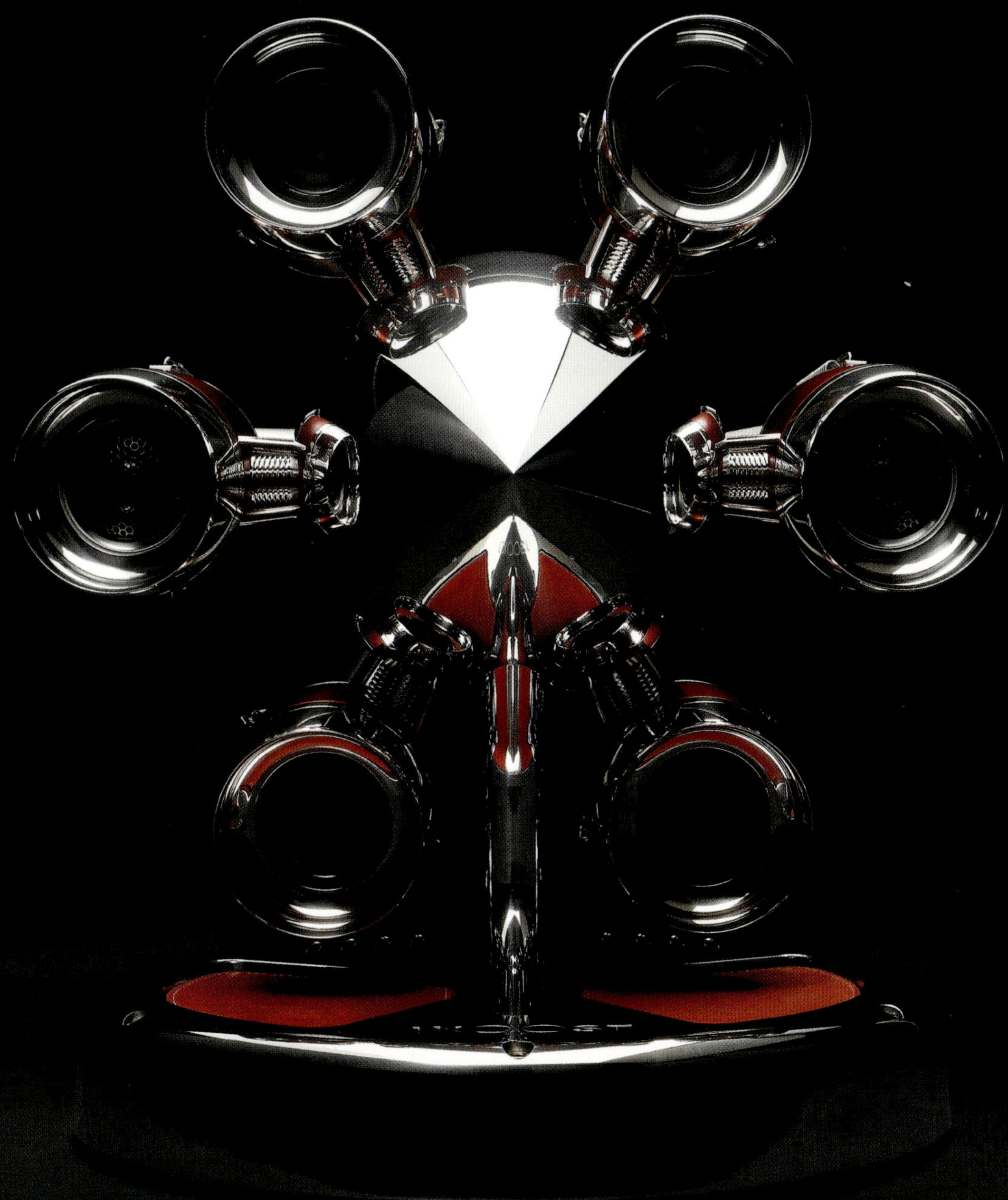

XiLO Scuderia

The XiLO Scuderia is one of the company's most exclusive iPod music player models. The limited-edition player incorporates parts from Ferrari automobiles and is handcrafted from wood, carbon fiber, and polished steel, giving it a high-quality industrial appearance. It can also be operated remotely via Bluetooth.

Janne Kytannen

Los Angeles, California

Finland-born Janne Kyttanen describes himself as a digital sculptor creating multidisciplinary work at the intersection of 3-D printing and virtual and augmented reality. His work is developing a commercial appeal. His furniture is exceptionally well-designed and the quality is superlative.

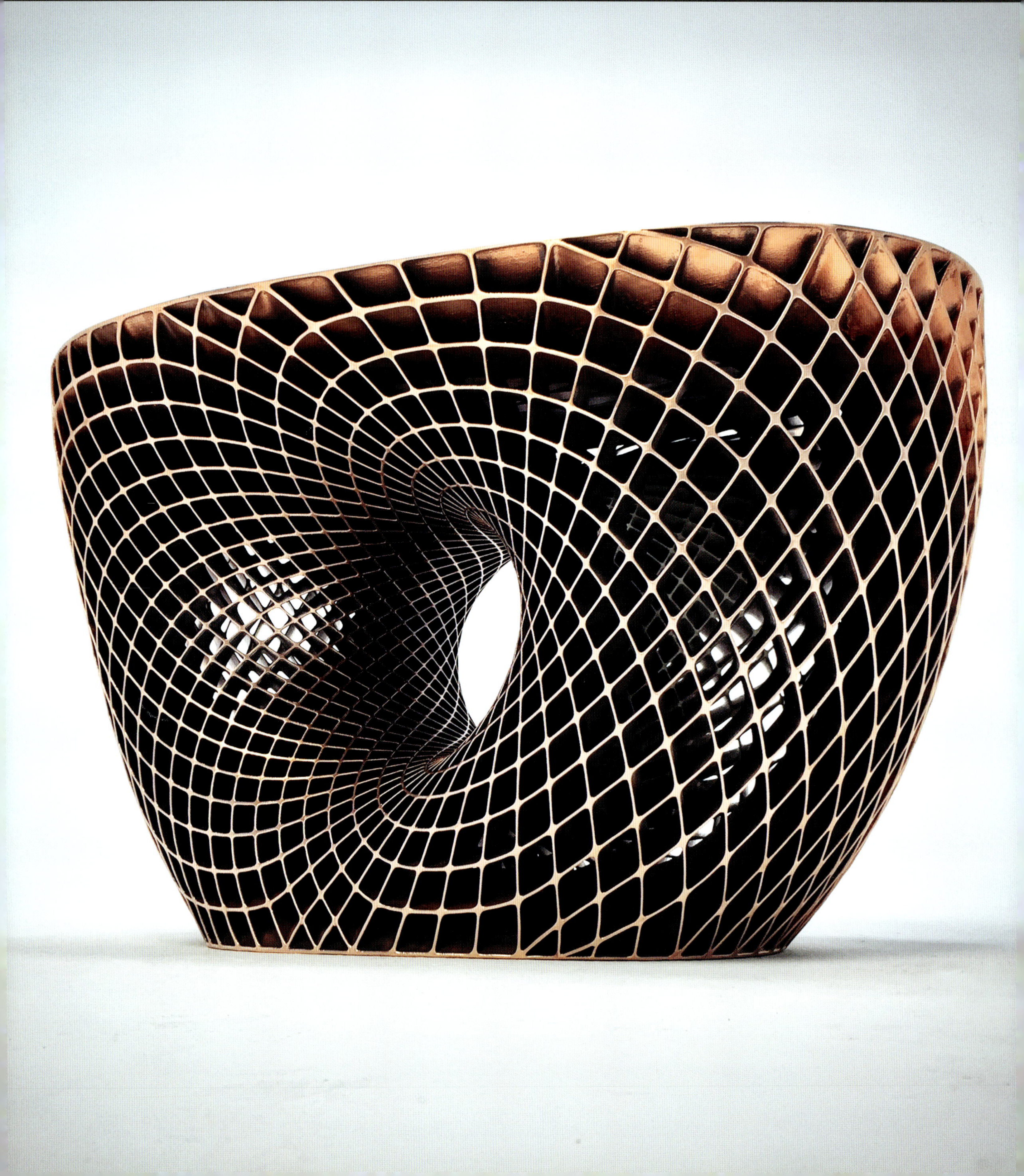

Avoid Stool

The bold Avoid stool was unveiled in 2014. Its intricate diamond-shaped structure was inspired by Lie sphere geometry, where the fundamental concept is the sphere. The stool is available in both bronze and 24-karat gold finishes.

Sedona Dining Table

The Sedona Dining Table was inspired by the red rocks of Sedona. With a hand-polished aluminum structure, this handsome piece suggests the peaks and plateaus of these formidable sandstone formations.

Sofa So Good

The ambitious Sofa So Good measures a sizable 150 mm × 75 mm × 55 mm. It is exceptionally lightweight at just 2.5 kg (5.5 pounds), but is incredibly strong due to the complex diamond pattern structure that will support up to 100 kg (220 pounds). The sofa's polished copper and chrome veneer gives it a luxurious appearance.

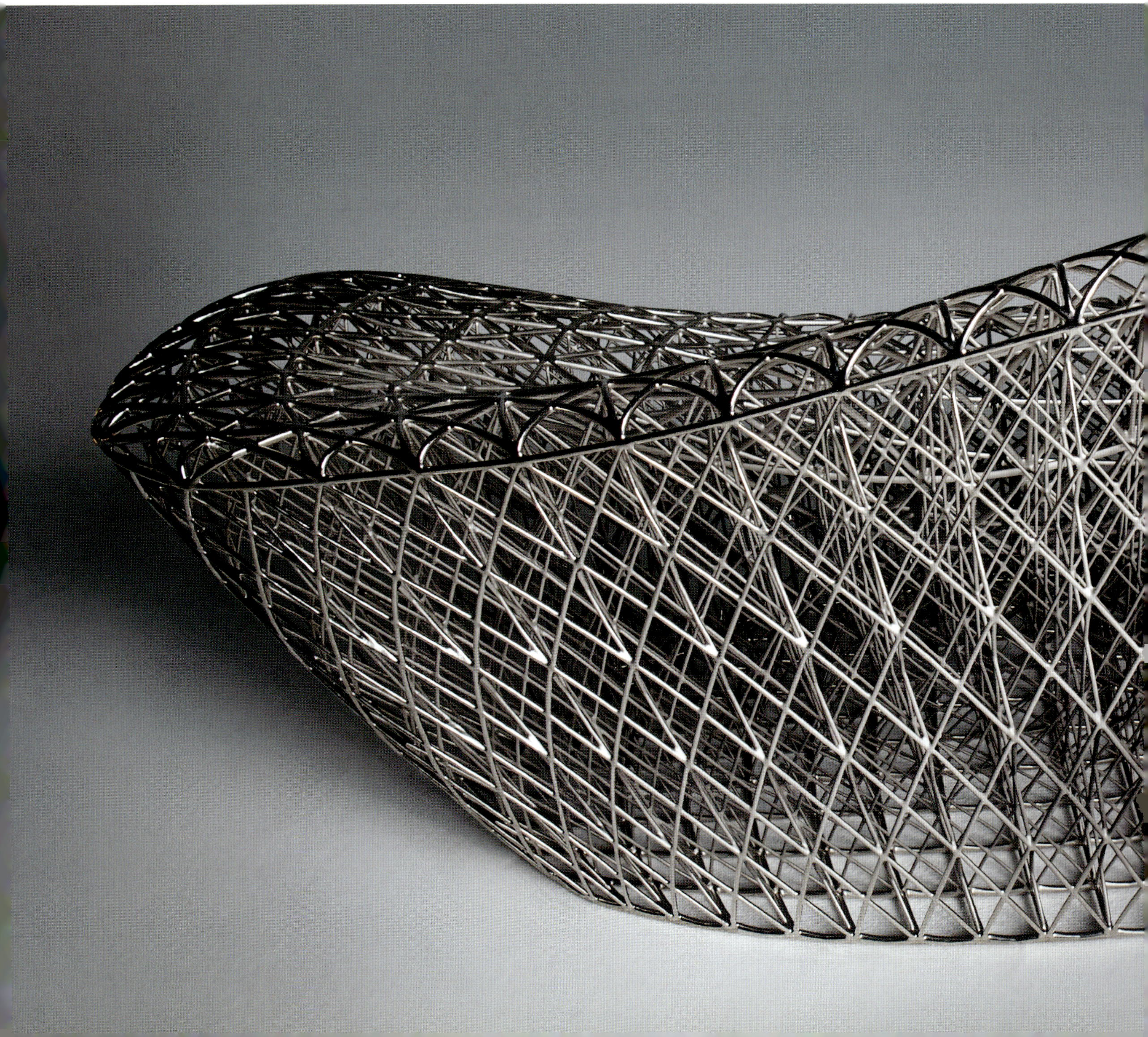

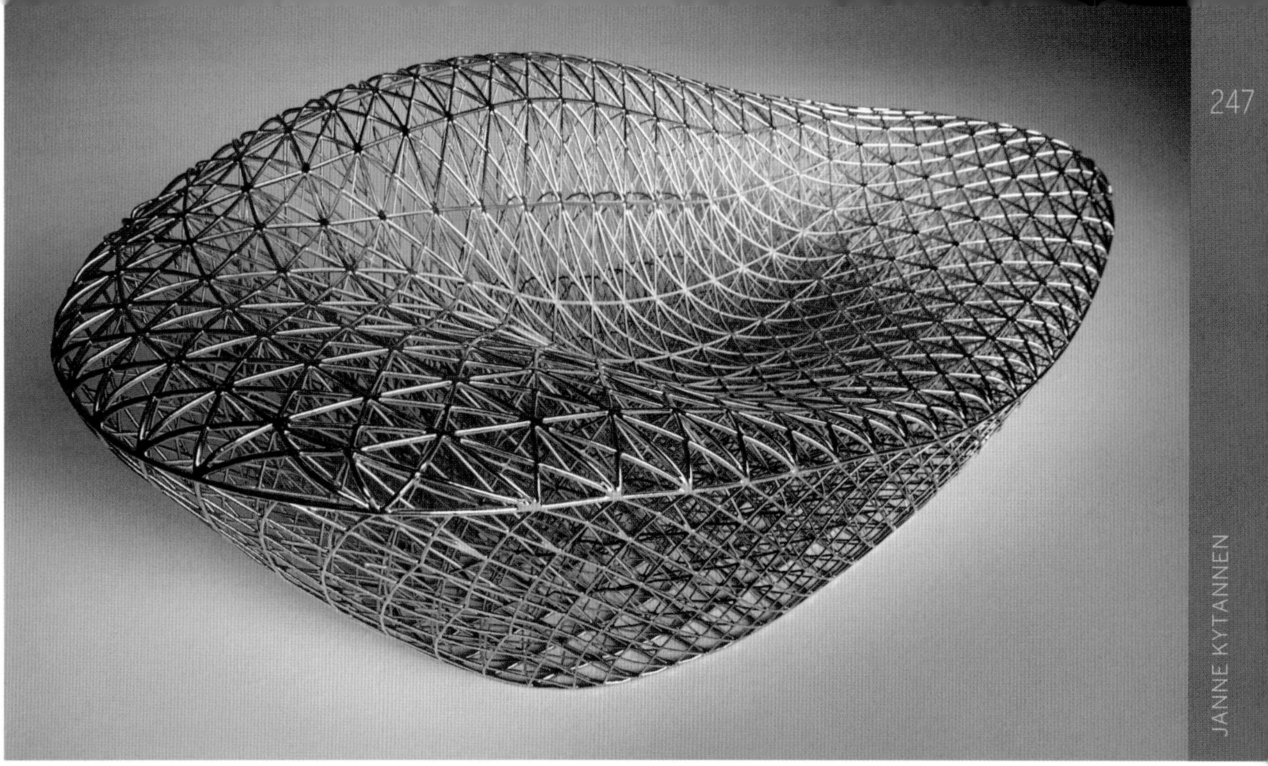

Marc Newson

London, England

Marc Newson is one of the most influential designers of his generation. Born in Sydney, Australia, he now resides in London. Over the course of his career, he has received many awards and his products are displayed in prominent global galleries.

Atmos 561

The Atmos 561 is a collaboration between Marc Newson and renowned watchmaker Jaeger-LeCoultre. This limited-edition clock (twenty-eight pieces) is a reinterpretation of Jaeger-LeCoultre's sublime 2008 Atmos Astronomique. A hand-blown Baccarat case encapsulates the complicated mechanism, and a sky chart depicts the Northern Hemisphere.

Lockheed Lounge

The curvaceous Lockheed Lounge was first unveiled in 1988, and took about six months to complete. It is fabricated from fiberglass, clad with sheets of highly polished aluminum, and held together with rivets, all of which give the piece a bold industrial appearance.

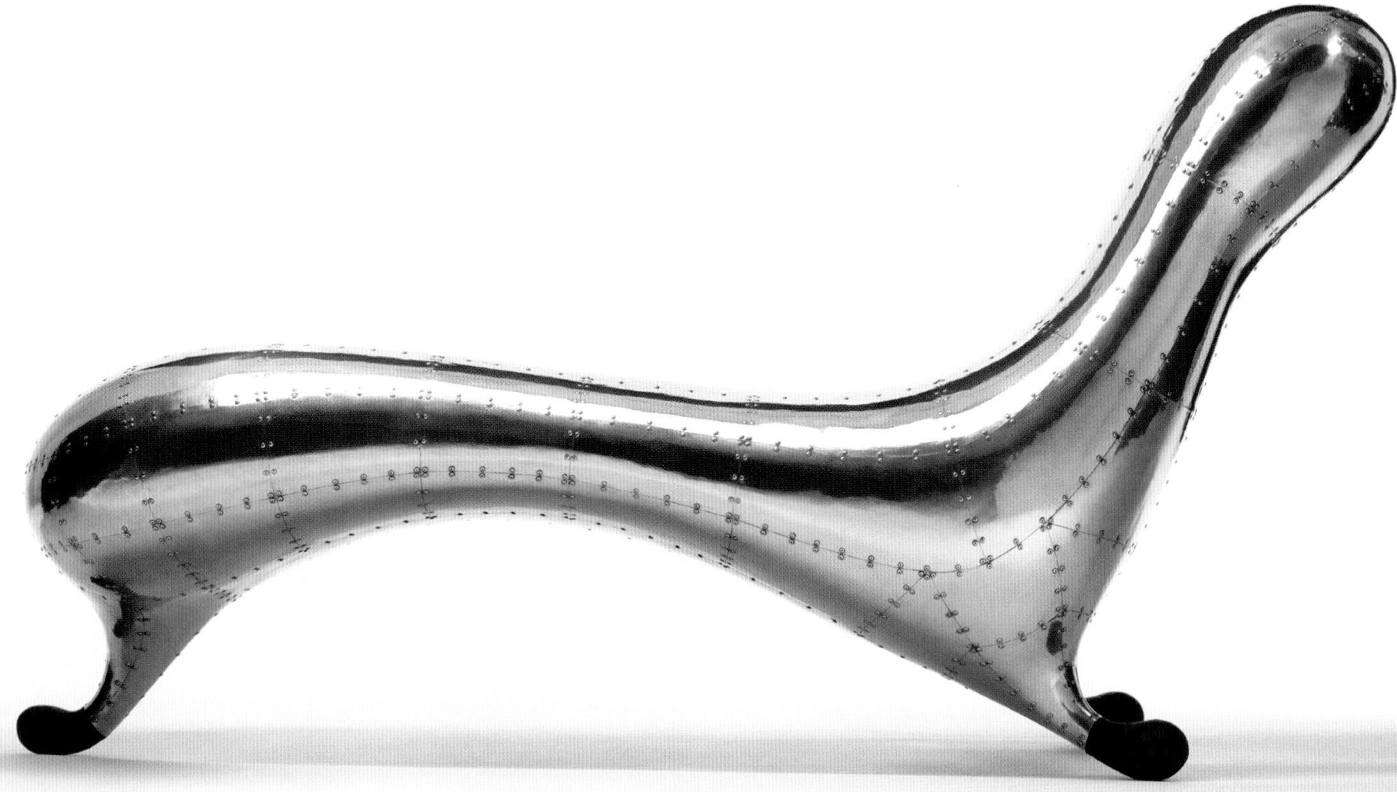

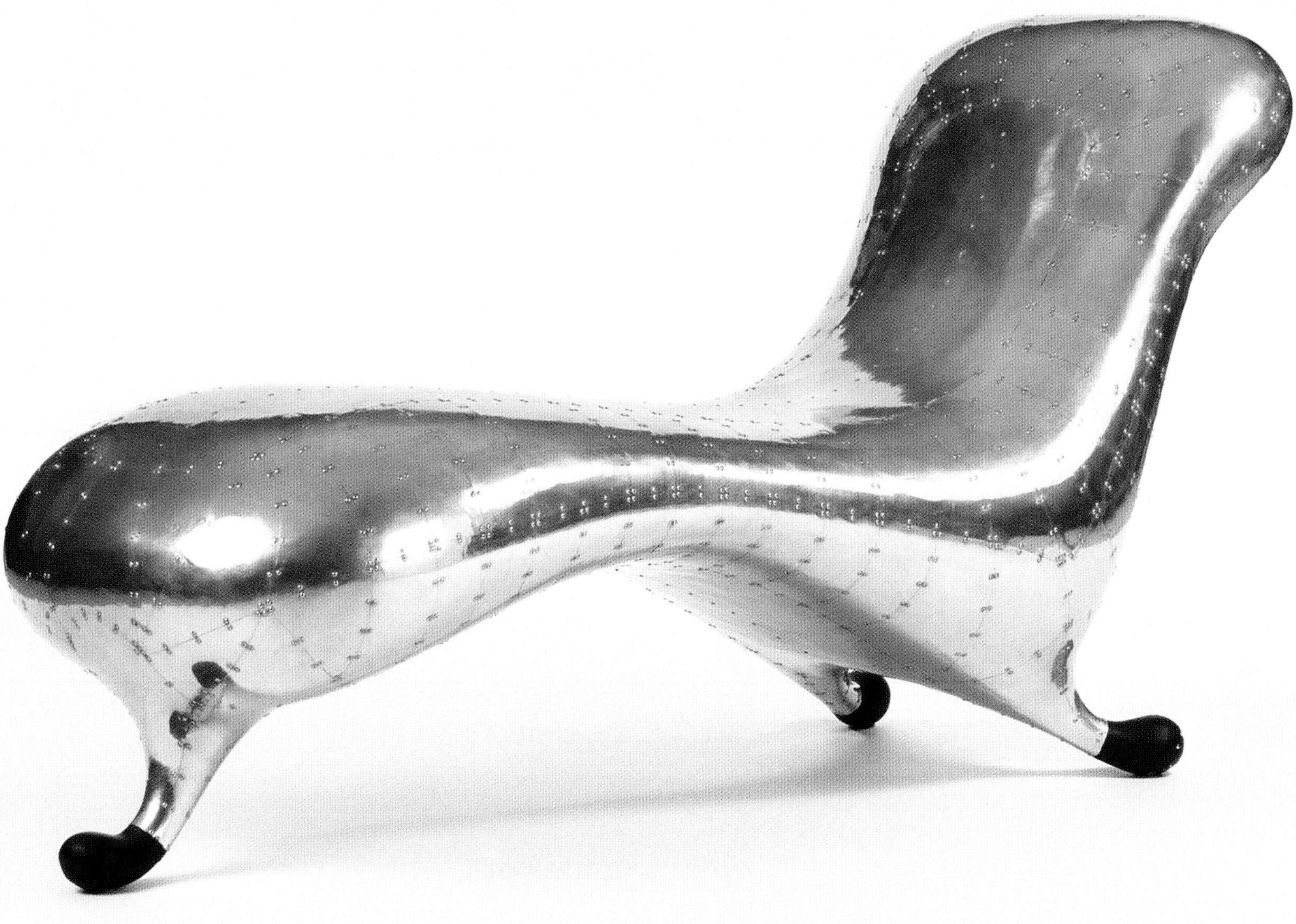

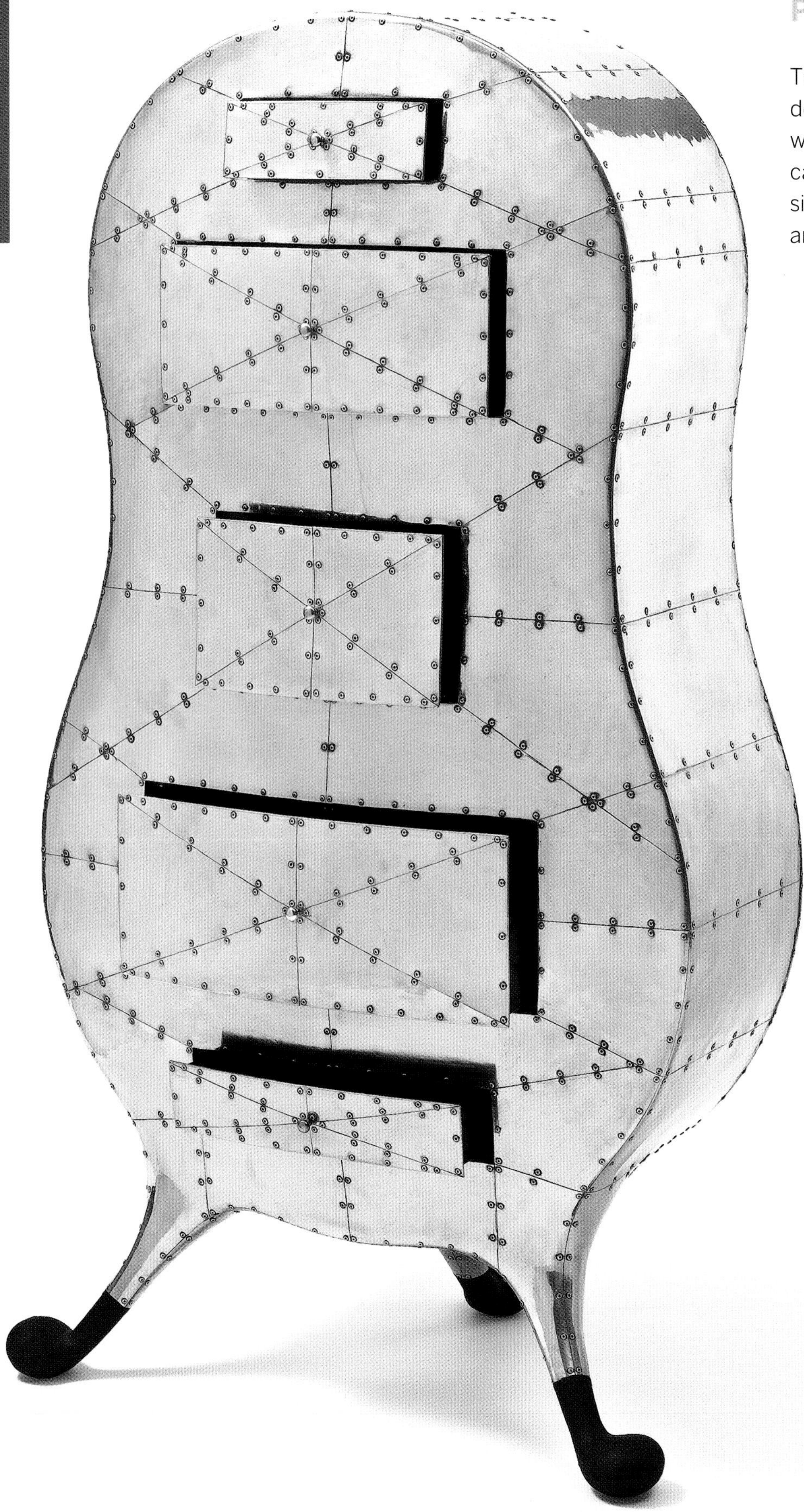

Pod of Drawers

The Pod of Drawers is one of Marc Newson's early designs, developed in 1987, and was inspired by works of legendary twentieth-century French cabinetmaker Andre Groult. Its construction is similar to the Lockheed Lounge. Sheets of aluminum are precision-wrapped around a fiberglass frame.

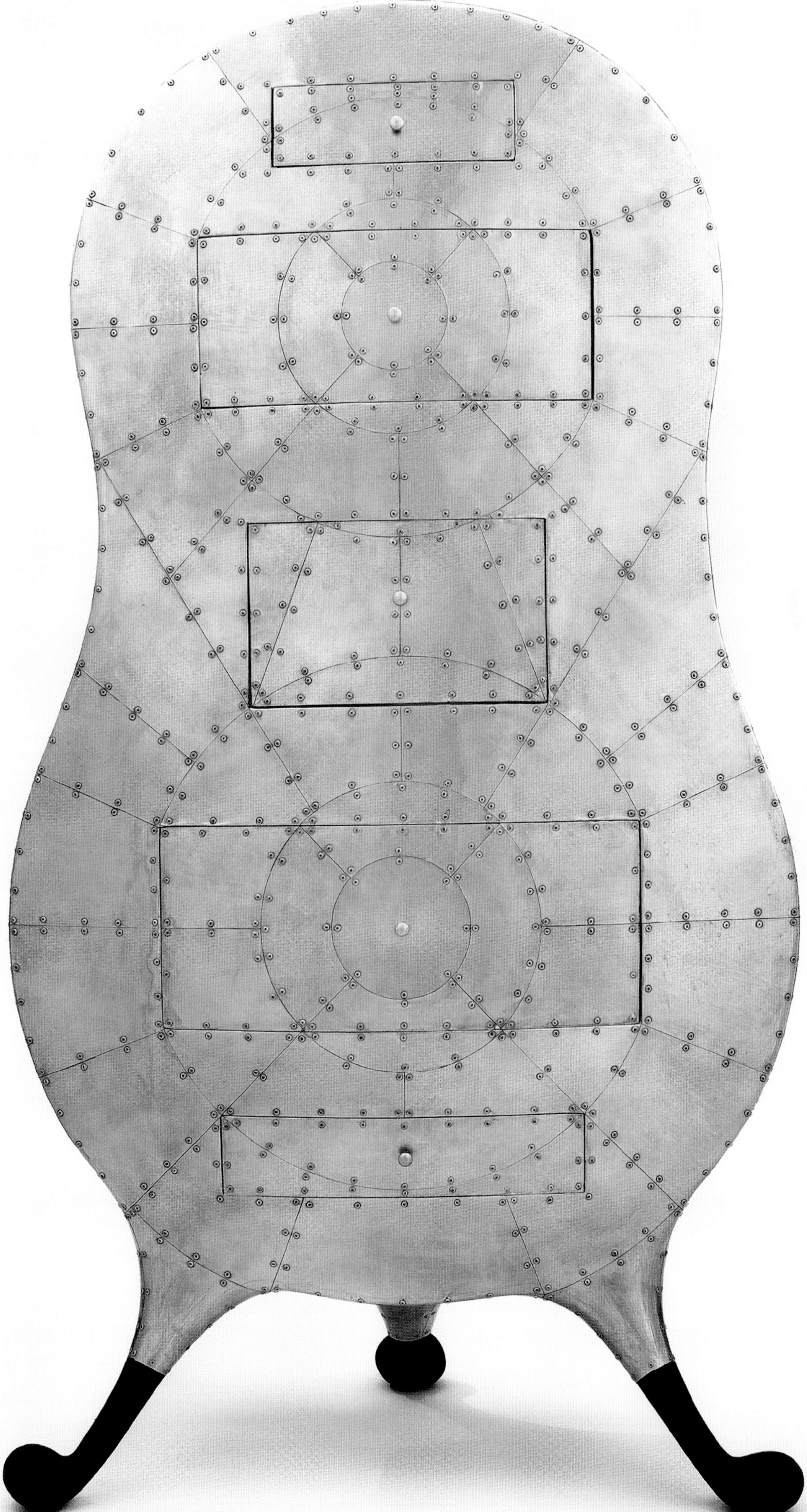

Metaxas

Amsterdam, The Netherlands

Metaxas, an exclusive audio equipment manufacturer in the Netherlands, was established in 1981, by music connoisseur Kostas Metaxas. The company has developed a range of luxury amplifiers that are handcrafted in small numbers and can be customized to the client's specifications.

Ikarus

The Ikarus is a futuristic-looking amplifier created for perfectionists. Over a thirty-five-year period, Kostas Metaxas has refined its design and performance. It is constructed from aluminum, and clients can choose the color and finish.

Opus

The Opus amplifier is handcrafted in Germany, and Kostas Metaxas claims it is manufactured to the same precision as a Patek Philippe watch. The clean, modernist lines of this piece of audio equipment would complement the living rooms of the world's finest homes.

module
muting

Soliloquy

The customizable Soliloquy has a unique, almost sculptural appearance. Certainly it looks unlike any other amplifier on the market. The elegant, curvaceous design pushes the limits on sound quality.

Peugeot Design Lab

Vélizy Villacoublay, France

Peugeot Design Lab has offices in Paris, Shanghai, and Sao Paolo. This allows the brand to devise innovative design solutions for companies around the world. Its products are constructed from the finest materials and most can be tailored to the client's precise specifications.

ONYX Chandelier

The ONYX chandelier is the second piece of furniture created by Peugeot Design Lab. It uses the latest 3-D printing technology combined with Banyan wood. Designed to make a bold statement, the chandelier can be personalized to the customer's exact specifications.

ONYX Sofa

The ONYX sofa may be the only seating constructed from carbon fiber and Volvic volcanic lava stone. At three meters long, this sofa also exudes massive proportions and had to be craned into the clients' living room. This might not be the world's most comfortable seating solution, or an affordable one. However, its dramatic industrial appearance makes a bold statement.

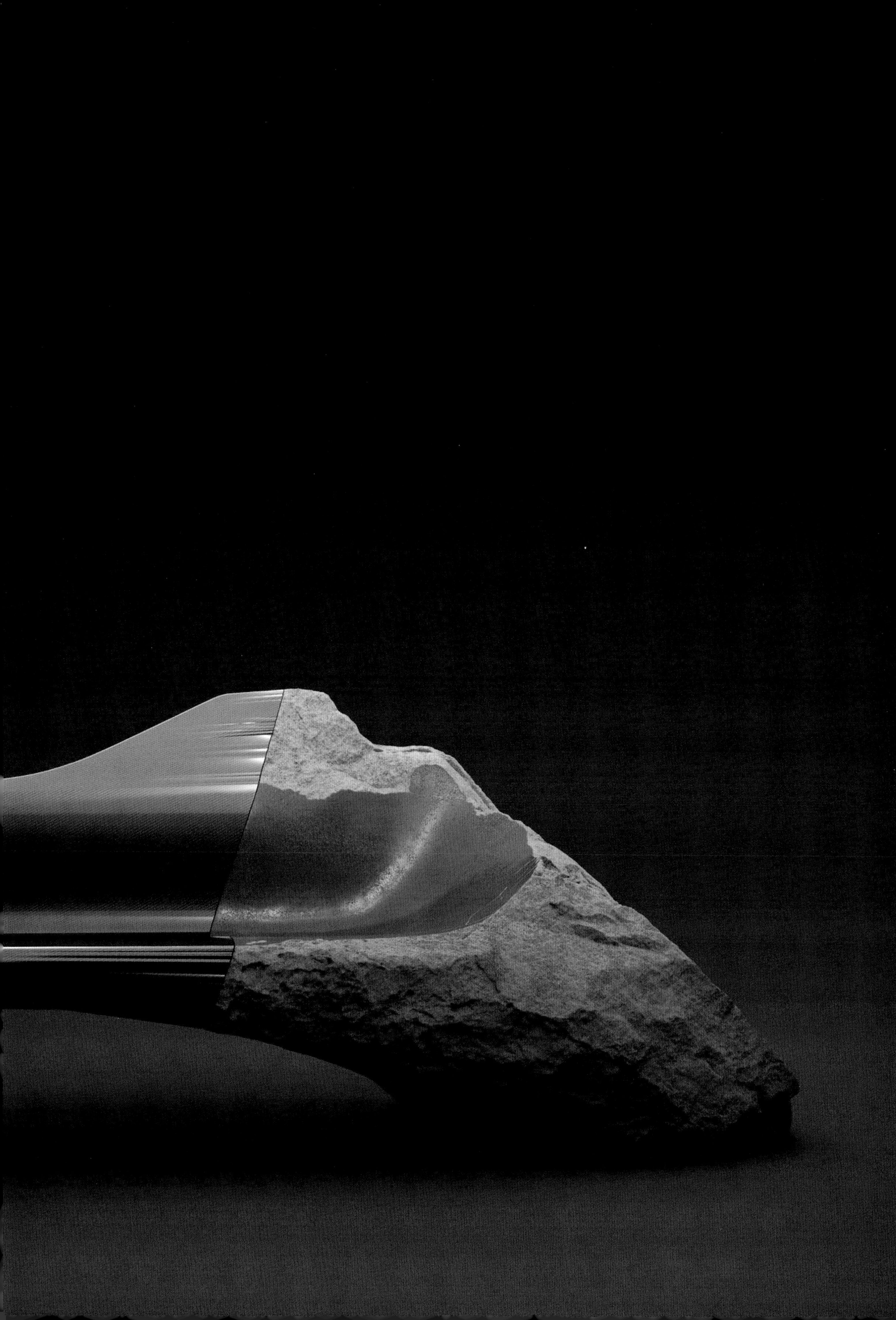

Piano

This unusual product is a collaboration between renowned piano manufacturer Pleyel and Peugeot Design Lab. The conventional three-leg format has been replaced by a single carbon fiber cantilever leg, giving the instrument a modernist architectural appearance. Other fine details include an ergonomic keyboard and black lacquered veneer.

Rolls Royce Lifestyle

Goodwood, England

The name Rolls Royce is associated with the automobile industry, but the company also produces a range of luxury accessories that complement its cars. These products display an exemplary level of craftsmanship and are produced from the finest-quality materials. They are manufactured in limited numbers to maintain exclusivity.

Cocktail Hamper

The Cocktail Hamper was designed by Sina Maria Eggl in collaboration with experts from the Dorchester London Hotel. The result is a beautifully handcrafted hamper fabricated from the finest leather and walnut. The materials are intended to replicate the interior of the company's marvelous cars.

Grand Tourer

The Grand Tourer from the Wraith Luggage Collection fits neatly into the boot compartment of a Rolls Royce automobile. This sensational suitcase is handcrafted from high-performance aluminium and carbon fiber that withstand the rigors of everyday life and maintain their good looks for many years. They can be personalized to the clients' exact specifications.

The Long Weekender

The Long Weekender from the Wraith Luggage Collection is the perfect traveling companion for the discerning motor enthusiast. This sublime, handcrafted item fits discreetly into the boot of your Rolls Royce. Each bag is fabricated from the finest-quality leather, features a unique magnetic zip, and is subtly monogrammed with the iconic company logo.

Splinter Works

Bath, England

Splinter Works is an exclusive British furniture, kitchen, and product manufacturer established in 2009 by progressive designers Miles Hartwell and Matt Withington. Their primary objective is to create a range of aspirational products. Their designs are highly experimental and use a broad range of top-quality materials.

Forum Ancient Oak

The Forum Ancient Oak is a modern twist on the traditional rustic kitchen. This product is constructed from twelve varieties of wood and has a beautiful organic appearance. Hand-laid applications of genuine white gold leaf give it a luxurious feel. Every detail has been meticulously executed by the finest German craftsmen, ensuring that it will last for years.

Metal-X²

The Metal-X² kitchen was awarded the highly coveted Best of Best award from Red Dot for its innovative "zeyko flybridge." The cantilevered, lacquered wood and metal island appears to float in mid-air.